METHODS IN MOLECULAR BIOLOGY

Series Editor
John M. Walker
School of Life and Medical Sciences
University of Hertfordshire
Hatfield, Hertfordshire, AL10 9AB, UK

For further volumes:
http://www.springer.com/series/7651

Skin Tissue Engineering

Methods and Protocols

Edited by

Sophie Böttcher-Haberzeth and Thomas Biedermann

University Children's Hospital Zurich, Zurich, Switzerland

Humana Press

Editors
Sophie Böttcher-Haberzeth
University Children's Hospital Zurich
Zurich, Switzerland

Thomas Biedermann
University Children's Hospital Zurich
Zurich, Switzerland

ISSN 1064-3745 ISSN 1940-6029 (electronic)
Methods in Molecular Biology
ISBN 978-1-4939-9472-4 ISBN 978-1-4939-9473-1 (eBook)
https://doi.org/10.1007/978-1-4939-9473-1

This Humana imprint is published by the registered company Springer Science+Business Media, LLC, part of Springer Nature.
The registered company address is: 233 Spring Street, New York, NY 10013, U.S.A.

Preface

Regenerative medicine and especially skin tissue engineering approaches are considered to be the most innovative advancements to the treatment of skin conditions and large wounds originating from traumas, burns, or other devastating skin diseases. The aim of this *Methods in Molecular Biology* volume is to provide protocols describing the isolation and culture of diverse cell types stemming from the skin and the use of these cells and cell constructs for wound healing, bioengineering applications, and translational medicine purposes. The chapters are divided into three sections describing the isolation and culture of diverse skin cells, managing these cells within co-culture systems and skin models, as well as using these skin models in a test setting.

This volume is addressed to basic and clinical researchers such as biologists, physicians, and biomedical engineers working with and being interested in basic science and clinically and laboratory applicable translational regenerative medicine, especially skin tissue engineering. In this respect, the chapters contain a detailed list of materials, step-by-step methodological descriptions, and, most importantly, a precious collection of personal advice from the authors to ensure success in their described technique.

Zurich, Switzerland

Sophie Böttcher-Haberzeth
Thomas Biedermann

Contents

Contributors

MARGARETE ARRAS • *Division of Surgical Research, University Hospital Zurich, University of Zurich, Zurich, Switzerland*
FRANÇOIS A. AUGER • *Centre de Recherche en Organogénèse Expérimentale de l'Université Laval/LOEX, Québec, QC, Canada; Division of Regenerative Medicine, CHU de Québec—Université Laval Research Center, Québec, QC, Canada; Faculty of Medicine, Department of Surgery, Université Laval, Québec, QC, Canada*
GERY BARMETTLER • *Center for Microscopy and Image Analysis, University of Zurich, Zurich, Switzerland*
CHARLES BASCOM • *Mason Business Centre, Procter & Gamble, Mason, OH, USA*
JENNIFER BOURLAND • *Centre de Recherche en Organogénèse Expérimentale de l'Université Laval/LOEX, Québec, QC, Canada; Division of Regenerative Medicine, CHU de Québec—Université Laval Research Center, Québec, QC, Canada; Faculty of Medicine, Department of Surgery, Université Laval, Québec, QC, Canada*
STEVEN T. BOYCE • *Research Department, Shriners Hospitals for Children, Cincinnati, OH, USA; Department of Surgery, University of Cincinnati, Cincinnati, OH, USA*
STEVEN BRADBURY • *Department of Biosciences, Durham University, Durham, UK*
MATTHIAS BRANDENBURGER • *Fraunhofer Research Institution for Marine Biotechnology and Cell Technology EMB, Lübeck, Germany*
LUDWIK K. BRANSKI • *Department of Surgery, University of Texas Medical Branch and Shriners Hospitals for Children®—Galveston, Galveston, TX, USA; Division of Hand, Plastic and Reconstructive Surgery, Department of Surgery, Medical University of Graz, Graz, Austria*
MURIEL CARIO • *INSERM 1035, University of Bordeaux, Bordeaux Cedex, France*
NIKOLA CESAROVIC • *Division of Surgical Research, University Hospital Zurich, University of Zurich, Zurich, Switzerland*
JINYING CHEN • *Department of Ophthalmology, The First Affiliated Hospital of Jinan University, Guangzhou, Guangdong, China*
SHUAI CHEN • *Department of General Surgery and Neonatal Surgery, Qilu Children's Hospital of Shandong University, Shandong, China*
ALVIN WEN CHOONG CHUA • *Department of Plastic, Reconstructive and Aesthetic Surgery, Singapore General Hospital, Singapore, Singapore*
SERGIO CORTEZ GHIO • *Centre de Recherche en Organogénèse Expérimentale de l'Université Laval/LOEX, Centre de Recherche du CHU de Québec-Université Laval, Axe Médecine Régénératrice, Québec, QC, Canada; Faculté de Médecine, Département de Chirurgie, Université Laval, Québec, QC, Canada*
LYDIA COSTELLO • *Department of Biosciences, Durham University, Durham, UK*
JÖRN DENGJEL • *Department of Biology, University of Fribourg, Fribourg, Switzerland; Department of Dermatology, Medical Center—University of Freiburg, Freiburg, Germany*
TERESA DICOLANDREA • *Mason Business Centre, Procter & Gamble, Mason, OH, USA*
THEA FLEISCHMANN • *Division of Surgical Research, University Hospital Zurich, University of Zurich, Zurich, Switzerland; Biological Central Laboratory, University of Zurich, Zurich, Switzerland*

JULIE FRADETTE • *Centre de Recherche en Organogénèse Expérimentale de l'Université Laval/LOEX, Québec, QC, Canada; Division of Regenerative Medicine, CHU de Québec—Université Laval Research Center, Québec, QC, Canada; Faculty of Medicine, Department of Surgery, Université Laval, Québec, QC, Canada*
NICOLA FULLARD • *Department of Biosciences, Durham University, Durham, UK*
LUCIE GERMAIN • *Centre de Recherche en Organogénèse Expérimentale de l'Université Laval/LOEX, Centre de Recherche du CHU de Québec-Université Laval, Axe Médecine Régénératrice, Québec, QC, Canada; Faculté de Médecine, Département de Chirurgie, Université Laval, Québec, QC, Canada*
ALFRED GUGERELL • *Department of Plastic and Reconstructive Surgery, Medical University of Vienna, Vienna, Austria*
MARTIN C. HARMSEN • *Department of Pathology and Medical Biology, University of Groningen and University Medical Centre of Groningen, Groningen, The Netherlands*
CRISTINA HAS • *Department of Dermatology, Medical Center—University of Freiburg, Faculty of Medicine, University of Freiburg, Freiburg, Germany*
YINGHONG HE • *Department of Dermatology, Medical Center—University of Freiburg, Faculty of Medicine, University of Freiburg, Freiburg, Germany*
ROBERT ISFORT • *Mason Business Centre, Procter & Gamble, Mason, OH, USA*
MAIKE KECK • *Department of Plastic and Reconstructive Surgery, Medical University of Vienna, Vienna, Austria; Department of Plastic and Reconstructive Surgery, Agaplesion Diakonieklinikum Hamburg, Hamburg, Germany*
MARTA A. KISIEL • *Occupational and Environmental Medicine, Medical Sciences, Uppsala University Hospital, Uppsala, Sweden*
AGNES S. KLAR • *Tissue Biology Research Unit, University Children's Hospital Zurich, University of Zurich, Zurich, Switzerland; Children's Research Center, University Children's Hospital Zurich, Zurich, Switzerland*
JOHANNA KOBER • *Department of Plastic and Reconstructive Surgery, Medical University of Vienna, Vienna, Austria*
CHARLI KRUSE • *Fraunhofer Research Institution for Marine Biotechnology and Cell Technology EMB, Lübeck, Germany; Institute for Medical and Marine Biotechnology, University of Lübeck, Lübeck, Germany*
SHYH-MING KUO • *Department of Biomedical Engineering, I-Shou University, Kaohsiung, Taiwan*
ANDREA L. LALLEY • *Research Department, Shriners Hospitals for Children, Cincinnati, OH, USA*
DANIELLE LAROUCHE • *Centre de Recherche en Organogénèse Expérimentale de l'Université Laval/LOEX, Centre de Recherche du CHU de Québec-Université Laval, Axe Médecine Régénératrice, Québec, QC, Canada; Faculté de Médecine, Département de Chirurgie, Université Laval, Québec, QC, Canada*
AMÉLIE LAVOIE • *Centre de Recherche en Organogénèse Expérimentale de l'Université Laval/LOEX, Centre de Recherche du CHU de Québec-Université Laval, Axe Médecine Régénératrice, Québec, QC, Canada; Faculté de Médecine, Département de Chirurgie, Université Laval, Québec, QC, Canada*
GAËTAN LE-BEL • *Centre de Recherche en Organogénèse Expérimentale de l'Université Laval/LOEX, Centre de Recherche du CHU de Québec-Université Laval, Axe Médecine Régénératrice, Québec, QC, Canada; Faculté de Médecine, Département de Chirurgie, Université Laval, Québec, QC, Canada*

SENG-TEIK LEE • *Department of Plastic, Reconstructive and Aesthetic Surgery, Singapore General Hospital, Singapore, Singapore*
SHIANG Y. LIM • *Department of Medicine, Surgery and Ophthalmology, University of Melbourne, East Melbourne, VIC, Australia; O'Brien Institute Department, St Vincent's Institute of Medical Research, Fitzroy, VIC, Australia*
MIRIAM LIPISKI • *Division of Surgical Research, University Hospital Zurich, University of Zurich, Zurich, Switzerland; Biological Central Laboratory, University of Zurich, Zurich, Switzerland*
GUEI-SHEUNG LIU • *Centre for Eye Research Australia, Royal Victorian Eye and Ear Hospital, East Melbourne, VIC, Australia; Department of Medicine, Surgery and Ophthalmology, University of Melbourne, East Melbourne, VIC, Australia; Department of Ophthalmology, The First Affiliated Hospital of Jinan University, Guangzhou, Guangdong, China; Menzies Institute for Medical Research, University of Tasmania, Hobart, TAS, Australia*
DONGRUI MA • *Department of Plastic, Reconstructive and Aesthetic Surgery, Singapore General Hospital, Singapore, Singapore*
HIROYUKI MATSUE • *Department of Dermatology, Graduate School of Medicine, Chiba University, Chiba, Japan; Medical Mycology Research Center, Chiba University, Chiba, Japan*
TAKAMITSU MATSUZAWA • *Department of Dermatology, Graduate School of Medicine, Chiba University, Chiba, Japan*
DOMINIQUE MAYRAND • *Centre de Recherche en Organogénèse Expérimentale de l'Université Laval/LOEX, Québec, QC, Canada; Division of Regenerative Medicine, CHU de Québec—Université Laval Research Center, Québec, QC, Canada; Faculty of Medicine, Department of Surgery, Université Laval, Québec, QC, Canada*
JUN MI • *Shandong Provincial Key Laboratory of Oral Tissue Regeneration and Laboratory for Tissue Engineering and Regeneration, School of Stomatology, Shandong University, Jinan, China; Cutaneous Biology Research Center, Massachusetts General Hospital/ Harvard Medical School, Boston, MA, USA*
GERALDINE M. MITCHELL • *Department of Medicine, Surgery and Ophthalmology, University of Melbourne, East Melbourne, VIC, Australia; O'Brien Institute Department, St Vincent's Institute of Medical Research, Fitzroy, VIC, Australia; Faculty of Health Sciences, Australian Catholic University, Melbourne, VIC, Australia*
VÉRONIQUE J. MOULIN • *Centre de Recherche en Organogénèse Expérimentale de l'Université Laval/LOEX, Québec, QC, Canada; Division of Regenerative Medicine, CHU de Québec—Université Laval Research Center, Québec, QC, Canada; Faculty of Medicine, Department of Surgery, Université Laval, Québec, QC, Canada*
YUUMI NAKAMURA • *Department of Dermatology, Graduate School of Medicine, Chiba University, Chiba, Japan*
MICHIYO NAKANO • *Department of Dermatology, Graduate School of Medicine, Chiba University, Chiba, Japan*
FLORA NICHOLLS • *Biological Central Laboratory, University of Zurich, Zurich, Switzerland*
AYAKO OIKAWA • *Department of Dermatology, Graduate School of Medicine, Chiba University, Chiba, Japan*
FILIPA PINTO • *Department of Molecular and Cell Biology, Boston University Henry M. Goldman School of Dental Medicine, Boston, MA, USA*

Daniel Popp • *Department of Surgery, University of Texas Medical Branch and Shriners Hospitals for Children®—Galveston, Galveston, TX, USA; Division of Hand, Plastic and Reconstructive Surgery, Department of Surgery, Medical University of Graz, Graz, Austria*

Stefan Przyborski • *Department of Biosciences, Durham University, Durham, UK; Reprocell Europe, Sedgefield, UK*

Mathilde Roger • *Department of Biosciences, Durham University, Durham, UK*

Makoto Senoo • *Department of Molecular and Cell Biology, Boston University Henry M. Goldman School of Dental Medicine, Boston, MA, USA*

Hieronymus P. Stevens • *Plastic Surgery Department, Velthuis Kliniek, Rotterdam, The Netherlands*

Daisuke Suzuki • *Department of Molecular and Cell Biology, Boston University Henry M. Goldman School of Dental Medicine, Boston, MA, USA*

Alain Taieb • *Department of Dermatology and Pediatric Dermatology, Bordeaux University Hospitals, INSERM U 1035, University of Bordeaux, Bordeaux, France*

Christian Tapking • *Department of Surgery, University of Texas Medical Branch and Shriners Hospitals for Children®—Galveston, Galveston, TX, USA; Department of Hand, Plastic and Reconstructive Surgery, Burn Trauma Center, BG Trauma Center Ludwigshafen, University of Heidelberg, Heidelberg, Germany*

Regine C. Tölle • *Department of Biology, University of Fribourg, Fribourg, Switzerland; Department of Dermatology, Medical Center—University of Freiburg, Freiburg, Germany*

Nathalie Tremblay • *Centre de Recherche en Organogénèse Expérimentale de l'Université Laval/LOEX, Québec, QC, Canada; Division of Regenerative Medicine, CHU de Québec—Université Laval Research Center, Québec, QC, Canada; Faculty of Medicine, Department of Surgery, Université Laval, Québec, QC, Canada*

Joris A. van Dongen • *Plastic Surgery Department, Velthuis Kliniek, Rotterdam, The Netherlands; Department of Pathology and Medical Biology, University of Groningen and University Medical Centre of Groningen, Groningen, The Netherlands*

Jiang-Hui Wang • *Centre for Eye Research Australia, Royal Victorian Eye and Ear Hospital, East Melbourne, VIC, Australia; Department of Medicine, Surgery and Ophthalmology, University of Melbourne, East Melbourne, VIC, Australia*

Jie Wen • *Shandong Provincial Key Laboratory of Oral Tissue Regeneration and Laboratory for Tissue Engineering and Regeneration, School of Stomatology, Shandong University, Jinan, China*

Xunwei Wu • *Shandong Provincial Key Laboratory of Oral Tissue Regeneration and Laboratory for Tissue Engineering and Regeneration, School of Stomatology, Shandong University, Jinan, China; Cutaneous Biology Research Center, Massachusetts General Hospital/Harvard Medical School, Boston, MA, USA*

Lin Xu • *Shandong Provincial Key Laboratory of Oral Tissue Regeneration and Laboratory for Tissue Engineering and Regeneration, School of Stomatology, Shandong University, Jinan, China; Department of Stomatology, Liaocheng People's Hospital, Shandong, China*

Xin Xu • *Shandong Provincial Key Laboratory of Oral Tissue Regeneration and Laboratory for Tissue Engineering and Regeneration, School of Stomatology, Shandong University, Jinan, China*

Urs Ziegler • *Center for Microscopy and Image Analysis, University of Zurich, Zurich, Switzerland*

Part I

Isolation and Culture of Skin Cells

Chapter 1

Isolation and Culture of Human Keratinocytes

Sergio Cortez Ghio, Gaëtan Le-Bel, Amélie Lavoie, Danielle Larouche, and Lucie Germain

Abstract

Culturing keratinocytes to form coherent epithelial tissue sheets has improved the treatment of extensively burned patients. Keratinocyte culture is also used to investigate various cellular and molecular mechanisms involved in different skin pathologies. To preserve stem cells during epithelial cell culture, reliable methods and conditions are of the utmost importance. Properly cultured keratinocytes will exhibit a consistent cuboid morphology and can proliferate for many passages. This chapter details materials needed and methods for all aspects of efficient keratinocyte culture for clinical applications, namely tissue sampling and transportation, isolation, routine culture, subculture, and cryopreservation.

Key words Cell therapy, Cultured epithelial autograft, Epidermis, Epithelium, Regenerative medicine, Tissue engineering, Wound healing

1 Introduction

Human epidermal cells (keratinocytes) have been cultured since 1957 [1]. Their massive in vitro expansion using feeder layers was then introduced by Rheinwald and Green in 1975 [2]. The identification of dispase to produce epidermal sheets [3] later resulted in the use of cultured autologous epidermal sheets to directly cover burn wounds or donor sites after autograft harvesting to facilitate skin regeneration [4, 5]. Keratinocyte cultures are also suitable to investigate various cellular and molecular mechanisms involved in different pathologies like skin cancer [6], epidermolysis bullosa [7], psoriasis [8], and amyotrophic lateral sclerosis [9] to name a few.

The therapeutic potential of a cultured epithelial autologous graft is associated with the ability of its cells to maintain their proliferative potential once implanted. One of the best ways to evaluate the quality of keratinocyte culture over time is to observe the morphological features of the cells by phase-contrast microscopy [10]. When cultured properly, keratinocytes exhibit a cuboid

Sophie Böttcher-Haberzeth and Thomas Biedermann (eds.), *Skin Tissue Engineering: Methods and Protocols*, Methods in Molecular Biology, vol. 1993, https://doi.org/10.1007/978-1-4939-9473-1_1,

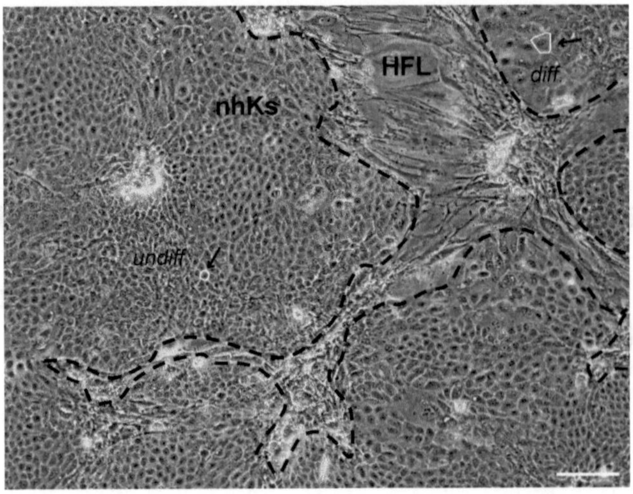

Fig. 1 First passage of a nhK-HFL co-culture at 10× under a phase-contrast microscope. The dashed black lines separate nhKs (colonies of smaller cuboidal cells) from the HFL (elongated cells). The nhK confluence is estimated at 75%. Undifferentiated nhKs ("undiff.") are smaller and have a greater nucleus/cytoplasm ratio than differentiated nhKs (directly under the "diff." annotation). Individual cells of each phenotype are identified with white contours and black arrows. Bar is 200 μm

morphology and form large colonies of small undifferentiated cells with a high nucleus/cytoplasm ratio referred to as holoclones. As passages advance and cells undergo differentiation, their proliferative potential declines. A larger proportion of keratinocytes then form small colonies of large cells with a low nucleus/cytoplasm ratio. These are called paraclones (abortive colonies). Completely differentiated keratinocytes display massively impaired proliferation capabilities and exhibit an elongated and irregular morphology (*see* Fig. 1).

In our lab, we use culture methods derived from those originally described by Rheinwald and Green to amplify keratinocytes for research and clinical applications [2]. Changes to the original protocols have indeed been made over time; human non-proliferative skin fibroblast feeder layers have now replaced murine embryonic immortalized fibroblast layers [11–13] and cholera toxin, a cyclic-AMP inducer, has been supplanted by isoproterenol, a widely available synthetic molecule which fills the same role in epithelial cells [14, 15]. Hence, this chapter details reliable methods and materials needed for all aspects of keratinocyte culture, namely tissue sampling and transport, isolation, routine culture, subculture, and cryopreservation. All methods described herein have been thoroughly tested and are used to treat patients [16, 17]. The products, reagents, and materials we use are selected based on availability, efficiency, quality, and safety according to the GMP guidelines (USP or GMP grades when available).

2 Materials

2.1 DME-Ham

DME-Ham: Combine three parts Dulbecco's modified Eagle's medium (DMEM) with one part Ham's medium in apyrogenic ultrapure water, add 3.07 g/L of $NaHCO_3$ (36.54 mM) and 24.3 mg/L of adenine (0.18 mM) solubilized in 312.5 μL/L of 2 N HCL, adjust pH to 7.1, sterilize by filtration through a 0.22 μm low-binding disposable filter (*see* Subheading 2.7), and store in the dark at 4 °C.

2.2 Sera

1. Fetal calf serum.
2. Fetal clone II serum.

Thaw sera at 4 °C or in cold water (*see* **Note 1**). Gently swirl it to resuspend its components. Inactivate in a 56 °C water bath for 30 min. To avoid repeated thawing and freezing cycles, distribute in single-use aliquots. Store at −20 °C or at −80 °C for long-term storage.

2.3 Additives

1. Insulin: 5 mg/mL in 5 mM HCl. Yields a 1000× stock solution.
2. Epidermal growth factor (EGF): 200 μg/mL in 10 mM HCl, dilute (1:20) the solution with DME-Ham (*see* Subheading 2.1) containing 10% v/v fetal clone II serum (*see* Subheading 2.2, **item 2**). Yields a 1000× stock solution.
3. Hydrocortisone: 5 mg/mL in 95% ethanol, dilute (1:25) the solution with DME-Ham (*see* Subheading 2.1). Yields a 500× stock solution.
4. Penicillin G/gentamicin: 50000 IU/mL Penicillin G and 12.5 mg/mL active gentamicin in apyrogenic ultrapure water. Yields a 500× stock solution.
5. Isoproterenol hydrochloride: Procure 0.2 mg/mL single use vials.
6. Fungizone: 0.25 mg/mL Amphotericin B in apyrogenic ultrapure water. Yields a 500× stock solution.

For all additives, except isoproterenol (*see* Subheading 2.3, **item 5**), sterilize by filtration through a 0.22 μm low-binding disposable filter (*see* Subheading 2.7, **item 1**). To avoid repeated thawing and freezing cycles, distribute in single-use aliquots. Store at −80 °C.

2.4 Complete Media

1. Tissue transport medium (tDMEM): Add 10% v/v of fetal calf serum (*see* Subheading 2.2, **item 1**) to high (4.5 g/L)-glucose DMEM (containing sodium pyruvate and L-glutamine), then add penicillin G/gentamicin and fungizone (dilute to 1×; *see* Subheading 2.3, **items 4** and **6**), and store in the dark at −20 °C to −80 °C for up to 6 months or at 4 °C for up to 10 days.

2. Complete keratinocyte culture medium (ckDME-Ham): Add 5% v/v of fetal clone II serum (*see* Subheading 2.2, **item 2**) to DME-Ham (*see* Subheading 2.1), and then add 1.06 mL/L of isoproterenol (*see* Subheading 2.3, **item 5**). Then, in this order, add insulin, epidermal growth factor, hydrocortisone, and penicillin G/gentamicin (dilute to 1×; *see* Subheading 2.3, **items 1–4**), and store in the dark at 4 °C for up to 10 days.
3. Cryopreservation medium: Add 10% v/v of dimethyl sulfoxide (DMSO) to fetal calf serum (*see* Subheading 2.2, **item 1**), keep on ice or store at 4 °C, and use within the day (*see* **Note 2**).

All frozen components can be thawed at 4 °C (*see* **Note 1**).

2.5 Other Solutions

1. Phosphate-buffered saline (PBS): 127 mM NaCl, 2.7 mM KCl, 6.5 mM Na_2HPO_4, and 1.5 mM KH_2PO_4 in apyrogenic ultrapure water, verify pH is between 7.35 and 7.45, and store at room temperature. Yields a 10× stock solution.
2. PBS—penicillin G/gentamicin/fungizone (PBS-P/G/F): Dilute 10× PBS (*see* Subheading 2.5, **item 1**) to 1× with apyrogenic ultrapure water, sterilize by filtration through a 0.22 μm low-binding disposable filter (*see* Subheading 2.7, **item 1**), and add penicillin G/gentamicin and fungizone (dilute to 1×; *see* Subheading 2.3, **items 4** and **6**).
3. HEPES/KCl/NaCl: 0.1 M 4-(2-Hydroxyethyl)-1-piperazineethanesulfonic acid (HEPES), 67 mM KCl, and 1.42 M NaCl in apyrogenic ultrapure water; adjust pH to 7.3; and store in the dark at 4 °C (*see* **Note 3**). Yields a 10× stock solution.
4. HEPES/KCl/NaCl—$CaCl_2$: Dilute 10× HEPES/KCl/NaCl to 1× (*see* Subheading 2.5, **item 3**) with apyrogenic ultrapure water, add 1 mM $CaCl_2$, adjust pH to 7.45, and store in the dark at 4 °C.
5. Thermolysin: 500 μg/mL in HEPES/KCl/NaCl—$CaCl_2$ (*see* Subheading 2.5, **item 4**), sterilize by filtration through a 0.22 μm low-binding disposable filter (*see* Subheading 2.7, **item 1**), store at 4 °C, and use within the day.
6. Trypsin/EDTA: 0.05% w/v Trypsin, 0.01% w/v EDTA, and 2.8 mM D-glucose in 1× PBS (*see* Subheading 2.5, **item 1**); add 100,000 IU/L of penicillin G, 25 mg/L active gentamicin, and 0.00075% v/v pre-sterile filtered 0.1% phenol red-water solution; adjust pH to 7.45; sterilize by filtration through a 0.22 μm low-binding disposable filter (*see* Subheading 2.7, **item 1**), to avoid repeated thawing and freezing cycles; distribute in single-use aliquots, and store at −20 °C to −80 °C.

2.6 Tissues and Cells

1. Normal human keratinocytes (nhK): Surgically removed 3–6 cm^2 skin biopsies.
2. Human fibroblast layer (HFL): Non-proliferative irradiated fibroblasts (*see* **Note 4**) extracted from normal skin biopsies (*see* **Note 4**) are seeded at 150,000 to 200,000/25 cm^2 into a culture flask (*see* Subheading 2.7, **item 9**) in 5 to 6 mL/25 cm^2 of ckDME-Ham (*see* Subheading 2.4, **item 2**). Medium must be changed every 7 days. HFL cultures are incubated in an 8% CO_2 and 100% humidity atmosphere at 37 °C. HFL cultures must be used within 30 days.

2.7 Labware

1. For volumes inferior to 100 mL: 0.22 μm Low-binding disposable filter. For volumes superior to 100 mL: filtration unit mounted with a 47 mm diameter and 0.22 μm filter set.
2. Sterile biopsy or urine collection container.
3. 50 mL Tubes.
4. 100 × 15 mm Cell culture Petri dishes.
5. Dissecting curved forceps.
6. Size 4 scalpel and size 22 blade.
7. Celstir® 50 mL suspension culture flask.
8. Parafilm® M.
9. 25 or 75 cm^2 tissue culture flasks.
10. Sterile cryogenic vials.
11. Freezing container.

3 Methods

Manipulations must be performed under a sterile laminar flow cabinet.

3.1 Tissue Sampling and Transport

1. Immediately following aseptic (*see* **Note 5**) surgical removal, put the skin biopsy into a sterile container filled with cold (4 °C) tDMEM.
2. Samples must be kept on ice and transported to a cell culture facility without delay.

3.2 Isolation

1. Wash the skin sample (*see* Subheading 3.1) by transferring it into a 50 mL tube containing 30 mL of PBS-P/G/F and agitating vigorously. With sterile forceps, transfer the sample into another 50 mL tube containing 30–45 mL of fresh PBS-P/G/F. Repeat the transfer eight additional times for a total of ten washes.

2. Use curved forceps to gently spread out the skin sample (epidermis facing top) in a Petri dish.
3. Use a scalpel with a size 22 blade to cut the sample into as many 3 mm × 100 to 150 mm strips as necessary. If the skin is hairy (i.e., scalp tissue), 3 mm × 3 mm strips are preferable.
4. Add 10 mL of cold (4 °C) thermolysin into the Petri dish containing the small skin strips (*see* **Note 6**).
5. Seal the Petri dish with Parafilm® M and incubate overnight at 4 °C.
6. Use two curved forceps to carefully separate the epidermis from the dermis (delicately pull hair if present to collect hair follicle keratinocytes). To dissociate nhKs, put the epithelial strips (epidermis and hairs) in a Celstir® suspension culture flask containing 20 mL warm (22–37 °C) trypsin/EDTA.
7. Incubate under agitation for 15–30 min at 37 °C.
8. Add 10 mL of warm (22–37 °C) ckDME-Ham into the flask. Transfer the 20 mL nhK suspension into a 50 mL tube containing 10 mL of warm ckDME-Ham. Rinse the culture suspension flask with 10 mL warm ckDME-Ham and add it to the 50 mL tube.
9. Use an automated cell counter or a hemocytometer to count the nhKs. Count is expected to be approximately 3–4 million nhK/skin biopsy cm^2 and cell size should be between 6 and 14 μm.
10. Use trypan blue staining and a hemocytometer to estimate cell viability. Cell viability is expected to be greater than 80%.
11. Centrifuge (300 × *g*) the nhK suspension for 10 min at room temperature.
12. Remove the supernatant and resuspend nhKs at the desired concentration in warm ckDME-Ham.
13. Seed nhKs at 100,000/25 cm^2 or more (*see* **Note 7**) into a culture flask pre-seeded with a HFL. Seed directly into the culture medium. Total medium volume should not exceed 7 mL/25 cm^2. If HFL medium change is scheduled the same day nhKs are seeded (7 days since last HFL medium change), replace only half the HFL-conditioned medium with warm ckDME-Ham instead.

3.3 Culture

1. Incubate nhK-HFL co-cultures (*see* Subheading 3.2, **step 13**, and Subheading 3.6, **step 10**) in an 8% CO_2 and 100% humidity atmosphere at 37 °C.
2. Change the culture medium three times a week, every 2–3 days. Remove the medium from the culture flask. Replace it with warm (22–37 °C) ckDME-Ham.

3. Monitor nhK morphology and confluence (*see* **Note 8**) daily under a microscope.
4. When nhKs reach 75–95% confluence, subculture (*see* Subheading 3.4) or cryopreserve (*see* Subheading 3.5) them. Do not let nhKs reach 100% confluence.

3.4 Subculture (Passage)

1. Remove the culture medium (*see* Subheading 3.3).
2. Depending on culture flask size, swiftly rinse nhKs with either 1 or 2 mL (for either 25 or 75 cm^2 culture flasks, respectively) warm (22–37 °C) trypsin/EDTA. Remove it.
3. Depending on culture flask size, add either 3 or 8 mL (for either 25 or 75 cm^2 culture flasks, respectively) of warm trypsin/EDTA into the culture flask.
4. Incubate at 37 °C until all nhKs are completely detached from the flask (verify cell detachment under a microscope). Time for complete detachment should be around 10 min. Do not incubate for more than 15 min (*see* **Note 9**).
5. Depending on culture flask size, neutralize trypsin activity by adding either 3 or 8 mL of warm (22–37 °C) ckDME-Ham and swirl the nhK suspension.
6. Vigorously pipette the nhK suspension up and down at least ten times to ensure suspension homogeneity.
7. Transfer the nhK suspension into a 50 mL tube.
8. Thoroughly rinse the culture flask with 4 mL warm ckDME-Ham and transfer the suspension to the 50 mL tube.
9. Use an automated cell counter or a hematimeter to count nhKs.
10. Use trypan blue staining and a hemocytometer to estimate cell viability. Cell viability is expected to be greater than 95%.
11. Centrifuge (300 × *g*) the nhK suspension for 10 min at room temperature.
12. Remove the supernatant and resuspend nhKs at the desired concentration in warm ckDME-Ham.
13. Seed nhKs at 100,000/25 cm^2 or more (*see* **Note 7**) into a culture flask pre-seeded with a HFL. Seed directly into the culture medium. Total medium volume should not exceed 7 mL/25 cm^2. If HFL medium change is scheduled the same day nhKs are seeded (7 days since last HFL medium change), replace only half the HFL-conditioned medium with warm ckDME-Ham instead.

3.5 Cryopreservation

1. Fill a freezing container with 100% isopropyl alcohol and store it at 4 °C until cooled.
2. Follow **steps 1** through **11** from Subheading 3.4.

3. Remove the supernatant and resuspend nhKs at the desired concentration (max. 10 million/mL) in cryopreservation medium. Put the tube on ice (*see* **Note 2**).
4. Aliquot in cryogenic vials on ice.
5. Put the cryogenic vials in the freezing container (*see* Subheading 3.5, **step 1**).
6. Store the container overnight at −80 °C. In these conditions, cell temperature should drop 1 °C/min.
7. Store cryogenic vials in liquid nitrogen for long-term storage.

3.6 Thawing

1. Put the cryogenic vial (*see* Subheading 3.5, **step** 7) in a 37 °C water bath. Do not let the nhK suspension thaw completely. A small ice pellet should remain.
2. Transfer the cell suspension from the cryogenic vial into a 50 mL tube containing 8–10 mL of cold (4 °C) ckDME-Ham.
3. Rinse the cryogenic vial with 1–1.5 mL of the suspension.
4. Centrifuge (300 × *g*) the nhK suspension for 10 min at room temperature.
5. Remove the supernatant and resuspend nhKs in 10 mL of warm (22–37 °C) ckDME-Ham.
6. Use an automated cell counter or a hemocytometer to count the nhKs.
7. Use trypan blue staining and a hemocytometer to estimate cell viability. Cell viability is expected to be greater than 80%.
8. Centrifuge (300 × *g*) the nhK suspension for 10 min at room temperature.
9. Remove the supernatant and resuspend nhKs at the desired concentration in warm ckDME-Ham (*see* **Note 10**).
10. Seed nhKs at 100,000/25 cm^2 or more (*see* **Note** 7) into a culture flask pre-seeded with a HFL. Seed directly into the culture medium. Total medium volume should not exceed 7 mL/25 cm^2. If HFL medium change is scheduled the same day nhKs are seeded (7 days since last HFL medium change), replace only half the HFL-conditioned medium with warm ckDME-Ham.

4 Notes

1. Sera and components can also be thawed more rapidly at room temperature or in a 37 °C water bath. However, we recommend that in that case they be diluted into culture media for imminent use and not refrozen.

2. DMSO is a toxic oxidative agent at temperatures above 10 °C. Work with cells in the presence of DMSO must be done quick and on ice.
3. HEPES undergoes degradation and becomes cytotoxic when exposed to fluorescent light [18].
4. (a) Other teams use alternative treatments to restrict fibroblast proliferation.

   (b) Fibroblasts can be cryopreserved in liquid nitrogen after irradiation (or preferred treatment).
5. The skin biopsy should be disinfected with 0.5% v/v chlorhexidine gluconate in isopropyl alcohol. Using any other disinfectant may compromise cell viability. Thoroughly rinse off the disinfectant from the skin biopsy with cold tDMEM. Failing to eliminate the disinfectant may also compromise cell viability.
6. Thermolysin allows for a better separation of the epidermis from the dermis, thus limiting fibroblast contamination in nhK cultures [19].
7. Seeding density is determined by the operator. nhK density is expected to double each day. Proliferation can vary from one population to another and decreases as passages advance. It is recommended to seed at 100,000 nhKs/cm^2 or more with an uncharacterized population. 95% Confluence should be reached within 5–7 days for the first few passages.
8. Here, we define confluence as the approximate surface percentage occupied by nhK colonies. It is estimated under a microscope by assessing how much of a given field of vision is occupied by nhK colonies (*see* Fig. 1). Do not take the HFL into account when evaluating confluence.

   Tips to evaluate confluence:

   (a) Technical replicates and multiple fields of the same flasks should be examined to increase accuracy.

   (b) When confluence is above 60%, it is easier to evaluate it by gaging how much surface is still available for nhKs to colonize.

   (c) If confluence varies greatly from field to field, seeding distribution is inadequate. Do not let more confluent areas differentiate for the sake of obtaining a higher mean confluence. However, note that trypsinizing nhKs under 75% confluence may result in count errors due to the overrepresented HFL.
9. For optimal trypsinization efficiency, do not stack culture flasks on top of each other in the incubator. Temperature is typically higher on the flask surface directly in contact with the incubator shelf.
10. A double wash is preferable to completely remove DMSO from the suspension (*see* **Note 2**).

Acknowledgments

The authors would like to thank the current and former members of the LOEX who have contributed to develop these protocols. This work was supported by the Canadian Institutes for Health Research (CIHR) grants MOP-12087 and FDN-143213 (L.G.), the Fondation des Pompiers du Québec pour les Grands Brûlés (FPQGB), the Fonds de Recherche du Québec-Santé (FRQS), and the Réseau de thérapie cellulaire et tissulaire (ThéCell) du FRQS. L.G. is the holder of a Tier 1 Canadian Research Chair from CIHR on Stem Cells and Tissue Engineering, and a Research Chair from the Fondation de l'Université Laval on Tissue-Engineered Organs and Translational Medicine.

References

1. Pruniéras M, Bonjean M, Chardonnet Y, Millet M (1957) Étude sur la trypsination de l'épiderme humain. Bull Soc Fr Dermatol Syphiligr 64:439
2. Rheinwatd JG, Green H (1975) Serial cultivation of strains of human epidermal keratinocytes: the formation of keratinizing colonies from single cells. Cell 6(3):331–343
3. Green H, Kehinde O, Thomas J (1979) Growth of cultured human epidermal cells into multiple epithelia suitable for grafting. Proc Natl Acad Sci 76(11):5665–5668
4. Gallico GG III, O'Connor NE, Compton CC, Kehinde O, Green H (1984) Permanent coverage of large burn wounds with autologous cultured human epithelium. N Engl J Med 311 (7):448–451
5. O'Connor N, Mulliken J, Banks-Schlegel S, Kehinde O, Green H (1981) Grafting of burns with cultured epithelium prepared from autologous epidermal cells. Lancet 317(8211):75–78
6. Purdie KJ, Lambert SR, Teh MT, Chaplin T, Molloy G, Raghavan M, Kelsell DP, Leigh IM, Harwood CA, Proby CM, Young BD (2007) Allelic imbalances and microdeletions affecting the PTPRD gene in cutaneous squamous cell carcinomas detected using single nucleotide polymorphism microarray analysis. Genes Chromosom Cancer 46(7):661–669
7. Bauer JW, Koller J, Murauer EM, De Rosa L, Enzo E, Carulli S, Bondanza S, Recchia A, Muss W, Diem A, Mayr E (2017) Closure of a large chronic wound through transplantation of gene-corrected epidermal stem cells. J Investig Dermatol 137(3):778–781
8. Piskin G, Sylva-Steenland RM, Bos JD, Teunissen MB (2006) In vitro and in situ expression of IL-23 by keratinocytes in healthy skin and psoriasis lesions: enhanced expression in psoriatic skin. J Immunol 176(3):1908–1915
9. Paré B, Touzel-Deschênes L, Lamontagne R, Lamarre MS, Scott FD, Khuong HT, Dion PA, Bouchard JP, Gould P, Rouleau GA, Dupré N, Berthod F, Gros-Louis F (2015) Early detection of structural abnormalities and cytoplasmic accumulation of TDP-43 in tissue-engineered skins derived from ALS patients. Acta Neuropathol Commun 3(1):5
10. Barrandon Y, Green H (1987) Three clonal types of keratinocyte with different capacities for multiplication. Proc Natl Acad Sci 84 (8):2302–2306
11. Braye F, Dumortier R, Bertin-Maghit M, Girbon JP, Tissot E, Damour O (2001) Cultured epidermis for the treatment of severe burns. A 2-year study (18 patients). Ann Chir Plast Esthet 46(6):599–606
12. Black AF, Bouez C, Perrier E, Schlotmann K, Chapuis F, Damour O (2005) Optimization and characterization of an engineered human skin equivalent. Tissue Eng 11(5–6):723–733
13. Bisson F, Rochefort É, Lavoie A, Larouche D, Zaniolo K, Simard-Bisson C, Damour O, Auger FA, Guérin SL, Germain L (2013) Irradiated human dermal fibroblasts are as efficient as mouse fibroblasts as a feeder layer to improve human epidermal cell culture lifespan. Int J Mol Sci 14(3):4684–4704
14. Green H (1978) Cyclic AMP in relation to proliferation of the epidermal cell: a new view. Cell 15(3):801–811
15. Ghoubay-Benallaoua D, Pécha F, Goldschmidt P, Fialaire-Legendre A, Chaumeil C, Laroche L, Borderie VM (2012) Effects of isoproterenol and cholera toxin on

human limbal epithelial cell cultures. Curr Eye Res 37(7):644–653

16. Larouche D, Cantin-Warren L, Desgagné M, Guignard R, Martel I, Ayoub A, Lavoie A, Gauvin R, Auger FA, Moulin VJ, Germain L (2016) Improved methods to produce tissue-engineered skin substitutes suitable for the permanent closure of full-thickness skin injuries. BioRes Open Access 5(1):320–329
17. Germain L, Larouche D, Nedelec B, Perreault I, Duranceau L, Bortoluzzi P, Beaudoin Cloutier C, Genest H, Caouette-Laberge-L, Dumas A, Bussière A, Boghossian E, Kanevsky J, Leclerc Y, Lee J, Nguyen MT, Bernier V, Knoppers BM, Moulin VJ, Auger A (2018) Autologous bilayered self-assembled skin substitutes (SASSs) as permanent grafts: a case series of 14 severely burned patients indicating clinical effectiveness. Eur Cell Mater 36:128–141
18. Zigler JS, Lepe-Zuniga JL, Vistica B, Gery I (1985) Analysis of the cytotoxic effects of light-exposed HEPES-containing culture medium. In Vitro Cell Dev Biol 21(5):282–287
19. Germain L, Rouabhia M, Guignard R, Carrier L, Bouvard V, Auger FA (1993) Improvement of human keratinocyte isolation and culture using thermolysin. Burns 19(2):99–104

Chapter 2

Promotion of Human Epidermal Keratinocyte Expansion in Feeder Cell Co-culture

Daisuke Suzuki, Filipa Pinto, and Makoto Senoo

Abstract

Co-culture of human epidermal keratinocytes with mouse 3T3-J2 feeder cells, developed by Green and colleagues, has been used worldwide to generate skin autografts since the early 1980s. In addition, co-culture with 3T3-J2 cells has served as a fundamental tool in skin stem cell biology as it allows the evaluation of self-renewal capacity of epidermal stem cells. This chapter describes a recent improvement in the Green method to promote further the expansion of human epidermal keratinocytes utilizing a small-molecule inhibitor of TGF-β signaling. This new protocol enables more rapid expansion of human epidermal keratinocytes in co-culture with not only 3T3-J2 cells but also other feeder cells including human dermal fibroblasts and human preadipocytes, two major alternatives to 3T3-J2 cells, with a long-term goal of developing customized skin autografts.

Key words Human keratinocytes, Skin transplantation, Feeder cell co-culture, TGF-β signaling, Small-molecule inhibitor

1 Introduction

Homeostasis of the epidermis is maintained by tissue-specific stem cells capable of self-renewal, proliferation, and differentiation [1, 2]. Culture of epidermal stem cells has contributed to the development of therapeutic strategies in regenerative medicine [3], including skin transplantation for patients with severe burn wounds. The Green method, developed for the expansion of autologous keratinocytes from a small skin biopsy, has enabled permanent wound closure with autografts in burn victims [3–6]. In addition, the Green method has been essential in studying skin stem cell biology as it allows the expansion of epithelial stem cells and the assessment of their self-renewal capacity [7–10].

The Green method utilizes 3T3-J2 cells, a unique mouse mesenchymal cell line, as a feeder layer to support the proliferative potential of epidermal progenitor cells long term [3, 7]. The

Sophie Böttcher-Haberzeth and Thomas Biedermann (eds.), *Skin Tissue Engineering: Methods and Protocols*, Methods in Molecular Biology, vol. 1993, https://doi.org/10.1007/978-1-4939-9473-1_2,

co-culture of human primary epidermal keratinocytes with 3T3-J2 cells produces cultured epidermal autografts (CEA). Since its inception in the 1980s, the CEA have been used worldwide for patients with severe burn wounds, often saving their lives [3–6]. However, patients with massive burn wounds can spare only limited skin donor sites and it requires several weeks of amplification of keratinocytes to prepare sufficient numbers of CEA. A methodology to promote the expansion of keratinocytes in Green's protocol would be beneficial as it can decrease hospitalization time for patients with massive burn injuries who undergo replacement of skin covering nearly the entire body surface area.

Transforming growth factor-β (TGF-β) signaling has been shown to suppress the growth of keratinocytes by inhibiting their cell cycle progression [11]. Effects of TGF-β signaling are mediated through a TGF-β receptor complex consisting of the type I TGF-β receptor (ALK5) and the type II TGF-β receptor (TGFBRII), both of which possess intrinsic serine/threonine kinase activity [12]. Upon binding of TGF-β ligands, TGFBRII activates ALK5 kinase by phosphorylation which, in turn, activates Smad2/3-mediated transcription of the genes required for the growth arrest of keratinocytes [13]. In co-culture with 3T3-J2 cells, there are three major sources of TGF-β ligands: keratinocytes [14], 3T3-J2 cells [15], and serum present in the co-culture media [16]. The presence of TGF-β ligands in co-culture could restrict the potential growth of human epidermal keratinocytes. Indeed, we have shown recently that TGF-β signaling is active in human keratinocytes in co-culture with 3T3-J2 cells as determined by the nuclear localization of Smad2/3 [17]. Accordingly, suppression of TGF-β signaling by RepSox, a cell-permeable small-molecule inhibitor of ALK5 kinase, increases the proliferation of human epidermal keratinocytes in co-culture with 3T3-J2 cells [17]. In addition, the use of TGF-β signaling inhibitors also enhances significantly the expansion of human epidermal keratinocytes in co-culture with other feeder cell types including human dermal fibroblasts and human preadipocytes, two major cell types utilized as alternatives to 3T3-J2 cells, with a long-term goal of developing customized skin autografts [18, 19].

In this chapter, we introduce an improved Green's method in which expansion of human epidermal keratinocytes is significantly enhanced by the use of pharmacological inhibition of TGF-β signaling.

2 Materials

2.1 Cells

1. Human primary epidermal keratinocytes (Cellntec) (*see* **Note 1**).
2. 3T3-J2 cells, derived from mouse embryonic fibroblasts [7], were provided by H. Green at Harvard Medical School.

3. PrimaPure human dermal fibroblasts (hDFs) (Genlantis): hDFs used in this protocol were derived from the dermis of normal human neonatal foreskin and have been verified to express fibroblast surface proteins (http://www.genlantis.com/human-fibroblasts.html).
4. Human preadipocytes (hPAs) were provided by M. Reilly at the University of Pennsylvania School of Medicine. hPAs used in this protocol were isolated from adipose tissues of patients who underwent plastic surgery in the abdominal region. The subjects were free of metabolic or endocrine diseases as assessed by routine clinical examination and laboratory tests.

2.2 Cell Culture

1. Dulbecco's Modified Eagle's Medium (DMEM): 4.5 g/L glucose, 584 mg/L L-glutamine, and 110 mg/L sodium pyruvate in DMEM.
2. Adenine (minimum 99%): 24 mg/mL in 0.2N HCl, sterilize with a 0.2 μm syringe filter. Store in 0.5 mL aliquots at −20°C.
3. T_3 (3,3′,5-triiodo-L-thyronine sodium salt) (minimum 95% HPLC): 10 μM in PBS (phosphate-buffered saline without calcium and magnesium), sterilize with a 0.2 μm syringe filter. Store in 0.5 mL aliquots at −20°C.
4. Hydrocortisone (>98%): 5 mg/mL in 95% EtOH. Store in 0.2 mL aliquots at −20°C.
5. Cholera toxin (approx. 95%): 0.1 mg/mL in dH_2O, sterilize with a 0.2 μm syringe filter. Store in 0.4 mL aliquots at −20°C.
6. Insulin from bovine pancreas: 0.1 mg/mL in 0.005 N HCl, sterilize with a 0.2 μm syringe filter. Store in 0.5 mL aliquots at −80°C.
7. Human epidermal growth factor (EGF): 20 μg/mL in 1% bovine serum albumin (BSA) aqueous solution, sterilize with a 0.2 μm syringe filter. Store in 0.5 mL aliquots at −80°C.
8. Basic fibroblast growth factor (bFGF): 10 μg/mL in 1% BSA aqueous solution. Store in 10 μL aliquots at −80°C.
9. Biotin: 1 mg/mL in 0.01 N NaOH, sterilize with a 0.2 μm syringe filter. Store in 0.1 mL aliquots at −80°C.
10. D-Pantothenic acid hemicalcium salt: 4 mg/mL in dH_2O, sterilize with a 0.2 μm syringe filter. Store in 0.1 mL aliquots at −80°C.
11. 2-[3-(6-Methylpyridine-2-yl)-1H-pyrazol-4-yl]-1,5-naphthyridine (RepSox, Selleck Chemicals): 25 mM in dimethyl sulfoxide (DMSO). Store in 50 μL aliquots at −80°C.

2.3 Other Reagents and Chemicals

1. Antibodies:
   (a) Rabbit anti-cytokeratin 5 (CK5) (polyclonal, BioLegend).
   (b) Mouse anti-p63 (4A4, Santa Cruz Biotechnology).

(c) Alexa 488-goat anti-mouse IgG (Molecular Probes).
(d) Alexa 488-goat anti-rabbit IgG (Molecular Probes).
(e) Alexa 594-goat anti-rabbit IgG (Molecular Probes).

2. 0.4% Trypan blue solution: Used to determine the viability of cells in cell counting.
3. 1% Triton X-100: 1% (v/v) Triton X-100 in PBS, store at room temperature.
4. 0.1% Tween-20 (PBS-T): 0.1% (v/v) Tween-20 in PBS, store at room temperature.
5. Blocking solution: 10% (v/v) fetal bovine serum (FBS) in PBS.
6. IntraSure kit (BD Biosciences): Used for intracellular staining.
7. 1% Rhodamine B: 1% (w/v) Rhodamine B in dH_2O, filter through Whatman filter papers, and store at room temperature.
8. Hoechst 33342: 1 mg/mL in DMSO as a stock solution. Store in 0.1 mL aliquots at −20 °C. To prepare working solution, dilute the stock solution to 1 μg/mL in PBS.

2.4 Instruments and Supplies

1. Humidified CO_2 incubators (5 and 10%).
2. CO_2 gas cylinders.
3. 37 °C Water bath.
4. Hemocytometer.
5. Inverted microscope.
6. Fluorescence microscope.
7. Flow cytometer.
8. γ-Irradiator with a Cesium-137 source.
9. Tissue culture plates (60, 100, 150 mm).
10. 15 and 50 mL polypropylene conical tubes.
11. 1.5 mL Eppendorf tubes.
12. Cryovials.
13. Syringe filter unit (0.2 μm).
14. Parafilm (Bemis Co. Ltd).

2.5 Cell Culture Media and Freezing Media

1. J2 medium: 100 U/mL penicillin, 100 μg/mL streptomycin, 10% (v/v) CS (bovine calf serum) in DMEM [7].
2. hPA medium: Mix DMEM and Ham's F-12 at a 1:1 ratio (v/v) and supplement with 10% Fetal FBS (v/v), 10 ng/mL EGF, 1 ng/mL bFGF, 8 μg/mL biotin, 4 μg/mL D-pantothenic acid, 8.7 μM insulin, 100 U/mL penicillin, and 100 μg/mL streptomycin [20].

3. FAD medium: Mix DMEM and Ham's F-12 at a 3:1 ratio (v/v) and supplement with 10% FBS (v/v), 100 U/mL penicillin, and 100 μg/mL streptomycin [7].
4. Complete FAD (cFAD) medium: 10 ng/mL EGF, 5 μg/mL insulin, 2×10^{-9} M 3,3′,5- triiodo-L-thyronine, 0.4 μg/mL hydrocortisone, 24 μg/mL adenine, and 1×10^{-10} M cholera toxin in FAD medium [7].
5. Freezing medium: Mix DMEM, serum (CS or FBS), and DMSO at a 4:5:1 ratio (v/v/v). To freeze 3T3-J2 cells and hDFs, use CS while FBS is used for human epidermal keratinocytes and hPAs.

3 Methods

3.1 Preparation of Feeder Cells

In this protocol, both 3T3-J2 cells and hDFs are expanded in J2 medium under 10% CO_2 atmosphere while hPAs are propagated in hPA medium under 5% CO_2 atmosphere. Keep in mind that the conditions of feeder cells, such as the viability, density, and passage number (*see* **Note 2**), are critical for successful expansion of human epidermal keratinocytes in feeder cell co-culture.

3.1.1 Culture, Passage, and Storage of 3T3-J2 Cells and hDFs

1. To expand 3T3-J2 cells and hDFs, one cryovial containing 1×10^6 cells is taken out of a liquid nitrogen tank and thawed in a 37 °C water bath for 3 min.
2. Spin down the cells at 200 × *g* for 5 min.
3. Remove freezing media and suspend the cells in 1 mL J2 medium.
4. Plate the resuspended cells in two 100 mm tissue culture plates that contain 10 mL J2 medium (5×10^5 cells per plate).
5. Place tissue culture plates in a humidified incubator with 10% CO_2 at 37 °C.
6. After overnight incubation (>16 h), replace media with 10 mL of fresh J2 medium. Culture medium should be changed every 2–3 days thereafter until the cells become subconfluent (*see* **Note 3**).
7. When the cells become subconfluent, remove culture media and wash the cells twice in 5 mL PBS.
8. Add 2 mL 0.25% trypsin-EDTA to the cells and incubate the plates in a humidified incubator with 10% CO_2 at 37 °C for 5 min.
9. Dissociate the cells by gentle pipetting, collect them in a 15 mL polypropylene tube, and neutralize the enzymatic reaction by adding the same volume (2 mL) of J2 medium.

10. Centrifuge the cells at 200 × *g* for 5 min.
11. Remove supernatant and resuspend the cells in 1 mL J2 medium.
12. Count cell numbers. Approximately 3 × 10^6 cells are collected from one subconfluent 100 mm tissue culture plate.
13. Replate 3T3-J2 cells or hDFs at a 1:10–1:15 ratio at the density of 3.3 × 10^3 cells per cm^2 into 100 mm tissue culture plates that contain 10 mL J2 medium. These secondary cultures become subconfluent in 5–7 days.
14. To prepare frozen stock, resuspend the cell pellet in freezing medium at the density of 2 × 10^6 cells per mL and aliquot 0.5 mL per cryovial (1 × 10^6 cells per vial). Place the cryovials in −80°C overnight and transfer them to a liquid nitrogen tank for longer storage.

3.1.2 Culture, Passage, and Storage of hPAs

1. To expand hPAs, one cryovial containing 1 × 10^6 hPAs is taken out of a liquid nitrogen tank and thawed in a 37 °C water bath for 3 min.
2. Spin down the cells at 200 × *g* for 5 min.
3. Remove freezing media and suspend the cells in 1 mL hPA medium.
4. Plate 1 × 10^6 hPAs in one 100 mm tissue culture plate that contains 10 mL hPA medium.
5. Place the tissue culture plate in a humidified incubator with 5% CO_2 at 37 °C.
6. After overnight incubation (>16 h), replace media with 10 mL of hPA medium. Culture medium should be changed every 2–3 days thereafter until the cells become subconfluent.
7. When the cells become subconfluent, remove culture media and wash the cells twice in 5 mL PBS.
8. Add 2 mL 0.25% trypsin-EDTA to the cells and incubate the plates in a humidified incubator with 5% CO_2 at 37 °C for 5 min.
9. Dissociate the cells by gentle pipetting, collect them in a 15 mL polypropylene tube, and neutralize the enzymatic reaction by adding the same volume (2 mL) of hPA medium.
10. Centrifuge the cells at 200 × *g* for 5 min.
11. Remove supernatant and resuspend the cells in 1 mL hPA medium.
12. Count cell numbers.
13. Replate hPAs at a 1:10 ratio at the density of 3.3 × 10^3 cells per cm^2 into 100 mm tissue culture plates that contain 10 mL hPA medium. These secondary cultures become subconfluent in 7–10 days.

14. To prepare frozen stock, resuspend the cell pellet in freezing medium at the density of 2×10^6 cells per mL and aliquot 0.5 mL per cryovial (1×10^6 cells per vial). Place the cryovials in −80°C overnight and transfer them to a liquid nitrogen tank for longer storage.

3.2 Preparation of Human Epidermal Keratinocyte Stocks

Human epidermal keratinocytes used in this protocol were derived from neonatal foreskin and were purchased from Cellntec. The cells have been tested negative for hepatitis B, hepatitis C, and HIV-1, and are free of bacteria, fungi, and mycoplasma contamination. Cryovials containing 5×10^5 viable cells were supplied on dry ice. The cells can be used immediately or transferred to a liquid nitrogen tank for later use.

1. To grow human epidermal keratinocytes in plastic culture without feeder cells, one cryovial containing 5×10^5 cells is thawed in a 37 °C water bath for 3 min.
2. Spin down the cells at 200 × *g* for 5 min.
3. Remove freezing media and suspend the cells in 1 mL CnT-PR medium.
4. Plate 5×10^5 human epidermal keratinocytes in one 100 mm tissue culture plate that contains 10 mL CnT-PR medium.
5. Place the tissue culture plate in a humidified incubator with 5% CO_2 at 37 °C.
6. After overnight incubation (>16 h), replace media with 10 mL of fresh CnT-PR medium. Culture medium should be changed every 2–3 days thereafter until the cells become subconfluent (*see* **Note 4**).
7. When the cells become subconfluent, remove culture media and wash the cells twice in 5 mL PBS.
8. Add 2 mL of 0.25% trypsin-EDTA to the cells and incubate the plate in a humidified chamber with 5% CO_2 at 37 °C for 5 min (*see* **Note 5**).
9. Tap the tissue culture plate to dislodge the cells.
10. Dissociate the cells by gentle pipetting, collect them in a 15 mL polypropylene tube, and neutralize the enzymatic reaction by adding 0.2 mL FBS (equivalent to 10% the volume of trypsin used).
11. Centrifuge the cells at 200 × *g* for 5 min.
12. Remove supernatant, wash the cells twice in 5 mL PBS, and resuspend in 1 mL CnT-PR medium.
13. Count cell numbers. Approximately 3–5 × 10^6 cells are collected from one subconfluent 100 mm tissue culture plate.

14. Replate human keratinocytes at a 1:15–1:20 ratio at the density of 4×10^3 cells per cm^2 into 100 mm tissue culture plates that contain 10 mL CnT-PR medium. These secondary cultures become subconfluent in 5–7 days.
15. To prepare frozen stock, resuspend the cell pellet in keratinocyte freezing medium at the density of 2×10^6 cells per mL and aliquot 0.5 mL per cryovial (1×10^6 cells per vial). Place the cryovials in −80°C overnight and transfer them to a liquid nitrogen tank for longer storage.

3.3 Promotion of Human Epidermal Keratinocyte Expansion in Co-culture with Feeder Cells Using a TGF-β Signaling Inhibitor

3.3.1 Preparation of Feeder Cell Layers

Prior to seeding keratinocytes, feeder cells will be mitotically arrested by either lethal γ-irradiation or mitomycin C treatment (*see* **Note 6**) to prevent propagation of the feeder cells in subsequent passages. In this protocol, we describe the use of lethally γ-irradiated feeder cells on the scale of a 100 mm tissue culture plate as an example. When other sizes of cell culture vessels are used, the number of feeder cells seeded and volume of the cell culture media used should be adjusted proportionally based on the surface area of the culture vessels (*see* **Note 7**).

1. Follow the procedures described in Subheading 3.1 and prepare desired numbers of 100 mm tissue culture plates with subconfluent feeder cells.
2. When feeder cells become subconfluent, seal the culture plates with Parafilm to maintain sterility and γ-irradiate the cells at a dose of 60 Gy (6000 rads).
3. After irradiation, replace media with 10 mL of fresh feeder cell culture medium and place the plates in a humidified incubator until use (*see* **Note 8**).

3.3.2 Setting Up Co-culture

1. Prepare a minimum amount of cFAD medium that can be consumed in 1 week as it contains proteins and chemicals that may degrade once diluted in media.
2. Prior to seeding human keratinocytes, replace feeder cell culture media with 10 mL cFAD medium and place the plates in a humidified incubator with 5% CO_2 at 37 °C for 1–2 h.
3. By following the steps in Subheading 3.2, prepare human epidermal keratinocytes in suspension. Use cFAD instead of CnT-PR medium to suspend epidermal keratinocytes.
4. Count viable keratinocyte numbers with the aid of trypan blue solution.
5. Adjust the density of keratinocytes to 10^3–10^5 viable cells per mL by diluting in cFAD medium.
6. Seed the desired number of human keratinocytes onto feeder cell layers. When rapid confluency of keratinocytes is planned, 10^4–10^5 cells per 100 mm tissue culture plate can be seeded (*see*

Note 9). Seeding at a higher density than 10^5 cells per 100 mm plate is not recommended as it does not fully allow relatively immature keratinocytes to expand. When the determination of the proliferative capacity of keratinocytes at a clonal level is required, 10^2–10^4 cells per 100 mm plate should be seeded to permit clonal growth of individual keratinocytes (*see* **Note 10**).

7. To promote the expansion of human epidermal keratinocytes in feeder cell co-culture, add a predetermined dose (*see* Subheading 3.4.1 below) of RepSox in cFAD medium in **step 6** above. A range between 0.1 and 1 μM RepSox treatment significantly boosts the growth of human epidermal keratinocytes with high proliferative potential [17] (*see* **Note 11**).
8. cFAD medium containing RepSox should be replaced every 2–3 days until desired confluency of keratinocytes is achieved.

3.4 Evaluation of the Effects by RepSox Treatment

In this section, we describe the methods to evaluate the effects of RepSox on feeder cell co-culture on (a) the growth of human keratinocyte progenitor cells as determined by intracellular staining of cytokeratin 5 (CK5), a marker for basal epidermal cells [21]; (b) clonal expansion of human epidermal keratinocytes as determined by Rhodamine B staining; and (c) protein expression of p63, an essential transcription factor for the maintenance of the proliferative potential of epidermal stem cells [10].

3.4.1 Determining the Effective Doses of RepSox in Promoting the Growth of Human Epidermal Keratinocytes in Feeder Cell Co-culture

Optimum doses of RepSox that promote human epidermal keratinocyte growth in feeder cell co-culture vary from one feeder cell type to another. The optimum dose of RepSox or any other TGF-β signaling inhibitors should be determined for each feeder cell type prior to performing co-culture of keratinocytes.

1. Inoculate 5×10^3 human epidermal keratinocytes (*see* Subheading 3.2) onto γ-irradiated feeder cells in 60 mm tissue culture plates that contain 5 mL cFAD medium.
2. Add increasing concentration of RepSox to the culture media. Cell culture media are changed every 2–3 days.
3. At day 10, remove cell culture media from the plates and wash the cells twice in 5 mL PBS.
4. Add 2 mL 0.25% trypsin-EDTA to the cells and incubate the plates in a humidified incubator with 5% CO_2 at 37 °C for 10 min.
5. Dissociate the cells by gentle pipetting, collect them in 15 mL polypropylene tubes, and neutralize the enzymatic reaction by adding 0.2 mL FBS (10% the volume of trypsin used).
6. Centrifuge the cells at 200 × *g* for 5 min.

7. Remove supernatant and gently suspend the cells in 1 mL cFAD medium.
8. Count total cell numbers (*see* **Note 12**).
9. Resuspend 5–20 × 10^5 cells in 50 μL PBS in 1.5 mL Eppendorf tubes.
10. Add 100 μL Reagent A (supplied as a component of IntraSure kit) to each tube, vortex the tubes, and incubate for 5 min at room temperature.
11. Add 1 mL PBS to each tube and centrifuge the cells at 800 × *g* for 5 min.
12. Remove supernatant and gently suspend the cells in 50 μL Reagent B (IntraSure kit).
13. Add rabbit anti-CK5 polyclonal antibody (1:500 dilution) to the cells and incubate for 30 min at room temperature.
14. Add 1 mL PBS to each tube and centrifuge the cells at 800 × *g* for 5 min.
15. Remove supernatant and gently suspend the cells in 50 μL PBS.
16. Add Alexa 488-goat anti-rabbit IgG antibody (1:250 dilution) and incubate the cells for 30 min in the dark at room temperature.
17. Add 1 mL PBS to each tube and spin down the cells at 800 × *g* for 5 min.
18. Remove supernatant, gently suspend the cells in 1 mL PBS, and centrifuge the cells again.
19. Remove supernatant and gently suspend the cells in 0.3 mL PBS.
20. Determine the frequency of $CK5^+$ cells by a flow cytometer.
21. Calculate absolute $CK5^+$ cell numbers using the total cell numbers obtained in **step 8** and the $CK5^+$ cell frequency determined in **step 20**. Plot the results for each concentration of RepSox as shown in Fig. 1 (*see* also **Note 13**).

3.4.2 Visualization of Human Epidermal Keratinocyte Clones in Feeder Cell Co-culture by Rhodamine B Staining

1. Inoculate 1 × 10^3 human keratinocytes (*see* Subheading 3.2) onto γ-irradiated feeder cells in 60 mm tissue culture plates in 5 mL cFAD medium in the presence or absence of predetermined doses of RepSox (*see* Subheading 3.4.1). Change cell culture media every 2–3 days.
2. At day 14, wash the cells twice in 2 mL PBS and fix them in 10% buffered formalin for 15 min at room temperature.
3. Remove formalin solution and wash the cells twice in 2 mL PBS.
4. Add 2 mL of 1% Rhodamine B solution to the cells and stain for more than 15 min at room temperature.

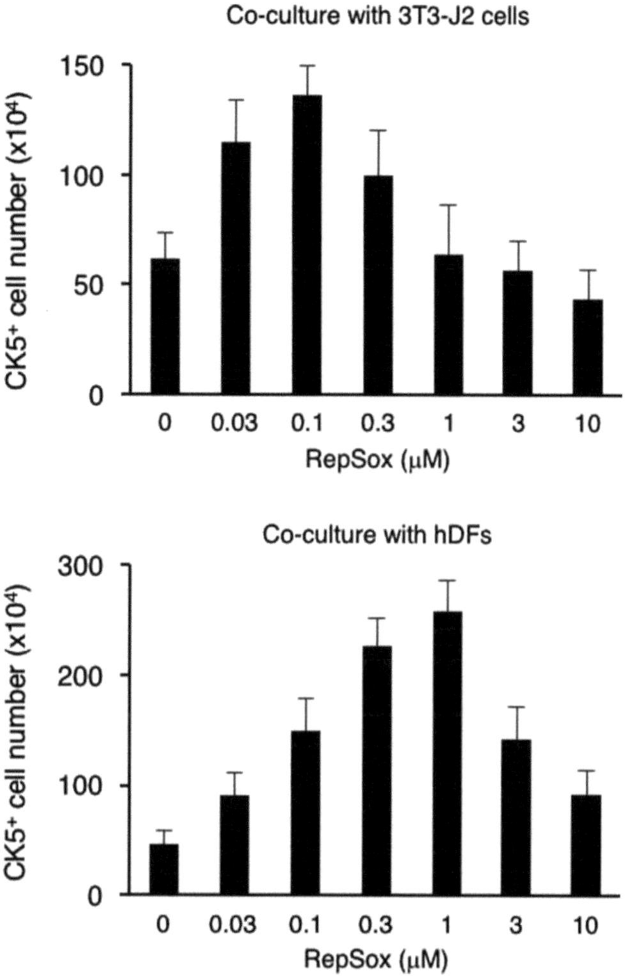

Fig. 1 Growth response of human epidermal keratinocytes to RepSox treatment in co-culture with feeder cells. Human epidermal keratinocytes were grown in co-culture with 3T3-J2 cells (upper) and hDFs (lower) in the absence or presence of increasing concentration of TGF-β signaling inhibitor RepSox as indicated. At day 10, the cells were harvested and the cell numbers were counted, followed by flow cytometric analysis of intracellular staining of cytokeratin 5 (CK5). $CK5^+$ epidermal cell numbers were plotted. Data shown are mean ± s.e.m. ($n = 3$)

5. Pour off Rhodamine B solution into a separate container (*see* **Note 14**).
6. Gently rinse the cells with tap water.
7. Invert the plates to air-dry.
8. Take photographs of the stained plates (*see* Fig. 2) and determine the sizes of individual clones as needed using NIH ImageJ or Adobe Photoshop software [17].

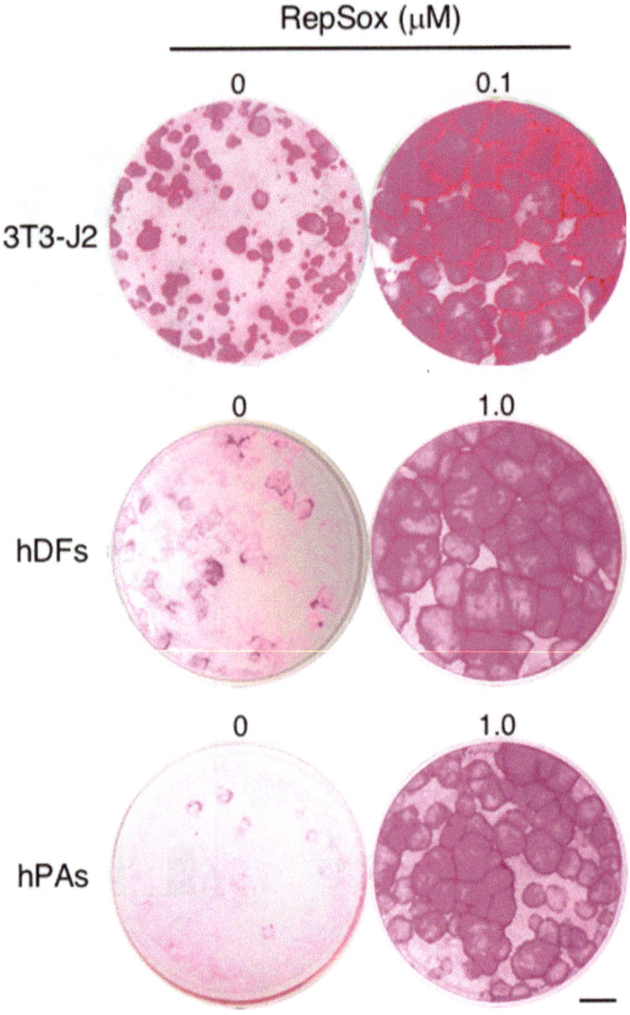

Fig. 2 Inhibition of TGF-β signaling promotes the expansion of human epidermal keratinocytes in co-culture with feeder cells. Data shown are representative images of Rhodamine B staining of human epidermal keratinocyte clones at day 14 of co-culture with 3T3-J2 cells (upper), hDFs (middle), and hPAs (lower) in the absence (left) or presence (right) of RepSox. Note that while 0.1 μM RepSox was used in co-culture with 3T3-J2 cells, 1.0 μM RepSox was used in co-culture with hDFs and hPAs. Equal numbers of human epidermal keratinocytes (1×10^3 cells) were plated in each well. Bar = 5 mm

3.4.3 Immunofluorescence Staining of p63 and CK5 in Human Epidermal Keratinocyte Clones in Feeder Cell Co-culture

1. Inoculate 1×10^3 human epidermal keratinocytes (*see* Subheading 3.2) onto γ-irradiated feeder cells in 60 mm tissue culture plates in 5 mL cFAD medium in the presence or absence of predetermined doses of RepSox (*see* Subheading 3.4.1). Change cell culture medium every 2–3 days.
2. At day 14, wash the cells twice in 2 mL PBS and fix them in 10% buffered formalin for 10 min at room temperature.

3. Remove formalin solution and add 2 mL of 1% Triton X-100 solution to permeabilize the cells for 5 min at room temperature.
4. Remove Triton X-100 solution and wash the cells three times in 2 mL PBS-T.
5. Add 2 mL blocking solution to the cells and incubate the plates for 30 min at room temperature.
6. Remove blocking solution and rinse the cells once in 2 mL PBS-T.
7. Add primary antibodies to the cells and incubate for 1 h at room temperature. Use a combination of rabbit anti-CK5 polyclonal antibody (1:2000) and mouse anti-p63 monoclonal antibody (1:1000), diluted in 2 mL PBS-T.
8. Remove the primary antibodies and wash the cells twice in 2 mL PBS-T for 5 min each.
9. Add secondary antibodies diluted in 2 mL PBS-T to the cells and incubate for 1 h in the dark at room temperature. Use a combination of Alexa 488-goat anti-mouse IgG and Alexa 594-goat anti-rabbit IgG (both 1:1000 dilution).
10. Remove the secondary antibodies and wash the cells three times in 2 mL PBS-T for 10 min each.
11. Add 2 mL of Hoechst 33342 solution to the cells and incubate for 5 min in the dark at room temperature.
12. Remove Hoechst 33342 and rinse the cells twice in 2 mL PBS-T for 5 min each.
13. Cover the cells with 2 mL PBS-T and analyze the signals using a fluorescence microscope as shown in Fig. 3.

4 Notes

1. Isolation of primary human epidermal keratinocytes has been described elsewhere. *See*, for example, reference [22].
2. We routinely expand 3T3-J2 cells at passages 4–7 and prepare a large number of frozen stocks before initiating epidermal cell co-culture. We typically use 3T3-J2 cells in co-culture at passages 6–10.
3. Do not allow feeder cells to become confluent as the morphology of the cells is altered and the potential of feeder cells to support keratinocyte growth may decline.
4. Do not allow human keratinocytes to become confluent because confluent keratinocytes are difficult to dissociate by enzymatic digestion and they tend to differentiate prematurely in subsequent passages.

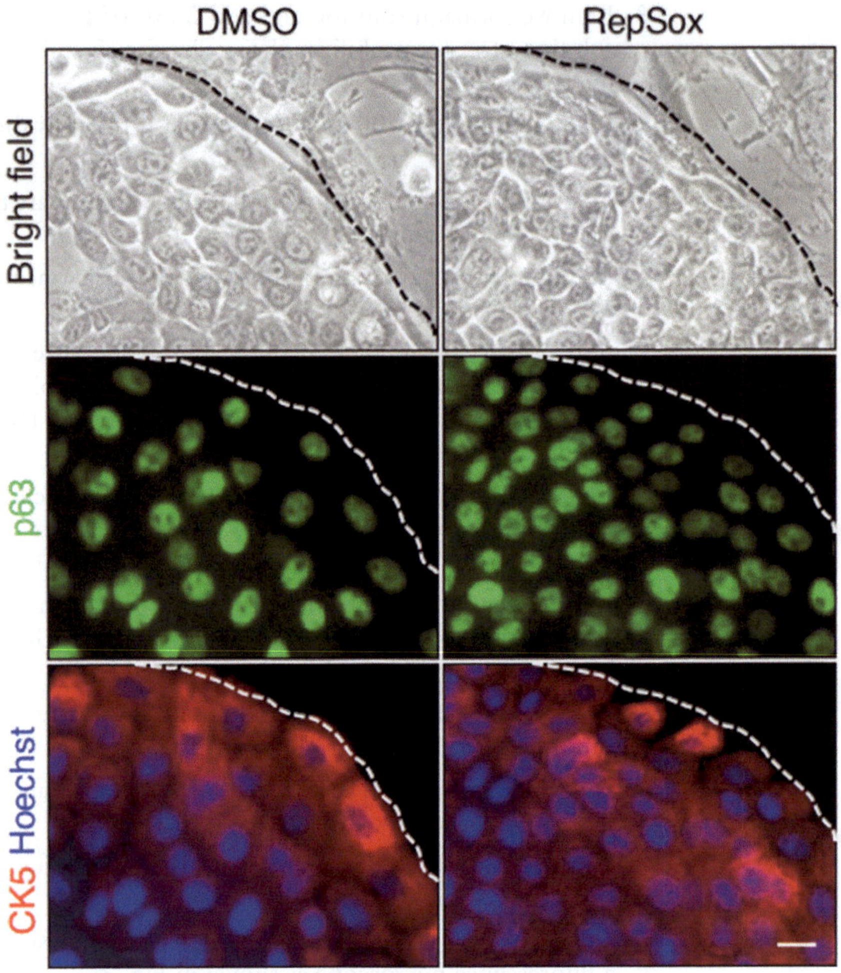

Fig. 3 Expression of p63 and CK5 in epidermal keratinocytes co-cultured with 3T3-J2 cells in the absence or presence of RepSox. Human epidermal keratinocytes were grown in co-culture with 3T3-J2 cells in the absence (left) or presence (right) of 0.1 μM RepSox for 14 days. Data shown are representative images of bright-field (upper) and immunofluorescence staining with anti-p63 antibodies (middle) and anti-CK5 antibodies and nuclear counterstaining with Hoechst 33342 (lower). Dotted lines indicate the clone borders. Bar = 10 μm

5. Accutase or TrypLE select can be used to more gently dissociate the cells compared with trypsin-EDTA.
6. Instead of γ-irradiation, mitomycin C treatment can be used to induce growth arrest in feeder cells as follows.
   (a) Add mitomycin C to feeder cell culture media in a 50 mL tube at a final concentration of 10 μg/mL.
   (b) Aspirate cell culture media from the plates and add 6 mL of freshly prepared mitomycin C-containing feeder cell media in each 100 mm plate.

Table 1
Number of feeder cells and the volume of cell culture media

Culture vessel	Diameter (mm)	Area (cm^2)	Volume of medium (mL)	Number of feeder cells
6-Well plate	35	9.6	3	5.0×10^5 cells
60 mm Plate	52	21	5	1.1×10^6 cells
100 mm Plate	84	55	10	3.0×10^6 cells

(c) Place the plates in a humidified incubator with 10% CO_2 at 37 °C for 2 h.

(d) Wash the cells three times in 5 mL PBS.

(e) Add 10 mL feeder cell culture media and return the plates to a humidified incubator with 10% CO_2 at 37 °C until use.

7. *See* Table 1 for expected numbers of subconfluent feeder cells at the time of keratinocyte inoculation and appropriate amounts of cell culture media used for feeder cell co-culture. When feeder cells are seeded at the density of 3.3×10^3 cells per cm^2, it will take 5–7 days for 3T3-J2 cells and hDFs to become subconfluent while it will require additional 2–3 days for hPAs to become subconfluent. Higher numbers of feeder cells can be seeded to obtain subconfluent feeder cell layers sooner. In such a circumstance, up to 2×10^6 viable cells per 100 mm plate can be seeded and the cells can be γ-irradiated in 1–2 days when the feeder cells become subconfluent and adhere tightly to the culture vessels.

8. Although feeder cells treated with γ-irradiation or mitomycin C can remain viable for a few weeks, we typically use them in 5–7 days after the treatment. Feeder cell media should be replaced every 2–3 days until use.

9. Keratinocytes grow faster when they are plated at higher densities, possibly due to an autocrine effect on cell proliferation [23].

10. At this plating density, small colonies comprised of 4–10 cells begin to form at days 4–5. By day 7, epidermal clones become macroscopically visible and they reach 5–15 mm in diameter by day 14. Cell culture vessels smaller than 6-well plates are not recommended for clonogenic analysis as keratinocyte colonies may fuse before they reach the size suitable for clonal evaluation.

11. Human epidermal keratinocytes expanded by RepSox treatment in feeder cell co-culture maintain the original high proliferative capacity as determined by serial passages in clonogenic culture with 3T3-J2 cells [17]. In addition, human epidermal keratinocytes expanded by RepSox treatment are capable of differentiation in response to Ca^{2+} stimulation [17].
12. Keratinocytes and feeder cells can be distinguished morphologically upon trypsin digestion: while keratinocytes are smaller in size with a smooth periphery, feeder cells are larger in size with an irregular shape. However, flow cytometric analysis with keratinocyte-specific markers is needed for more accurate quantification of keratinocyte numbers (*see* Fig. 1, for example).
13. Other widely used TGFBRI/ALK5 inhibitors, such as LY364947 and SB525334, also enhance the growth of human epidermal keratinocytes in co-culture with 3T3-J2 cells in a dose-dependent manner [17]. However, the effect of RepSox was superior to any other TGFBRI/ALK5 inhibitors tested [17]. In the case of co-culture with hPAs, we investigated the effects of RepSox at 0.1–1 μM and found that the dose-dependent stimulation of human keratinocyte growth was similar to that in co-culture with hDFs.
14. Rhodamine B solution can be reused multiple times. Used solution should be collected in a separate container and filtered through Whatman filter papers to remove debris before use.

Acknowledgments

The authors thank Jeff Holcombe for proofreading of the manuscript. This study was supported by an R01AR066755 grant from the National Institute of Arthritis and Musculoskeletal and Skin Diseases of the National Institute of Health to M.S.

References

1. Blanpain C, Fuchs E (2014) Stem cell plasticity. Plasticity of epithelial stem cells in tissue regeneration. Science 344:1242281. 1–10
2. Donati G, Watt FM (2015) Stem cell heterogeneity and plasticity in epithelia. Cell Stem Cell 16:465–476
3. Green H (2008) The birth of therapy with cultured cells. BioEssays 30:897–903
4. Chua AW, Khoo YC, Tan BK, Tan KC, Foo CL, Chong SJ (2016) Skin tissue engineering advances in severe burns: review and therapeutic applications. Burns Trauma 4:3
5. Mcheik JN, Barrault C, Levard G, Morel F, Bernard FX, Lecron JC (2014) Epidermal healing in burns: autologous keratinocyte transplantation as a standard procedure: update and perspective. Plast Reconstr Surg Glob Open 2:e218
6. Sun BK, Siprashvili Z, Khavari PA (2014) Advances in skin grafting and treatment of cutaneous wounds. Science 346:941–945
7. Rheinwald JG, Green H (1975) Serial cultivation of strains of human epidermal keratinocytes: the formation of keratinizing colonies from single cells. Cell 6:331–343
8. Barrandon Y, Green H (1987) Three clonal types of keratinocyte with different capacities

for multiplication. Proc Natl Acad Sci U S A 84:2302–2306

9. Oshima H, Rochat A, Kedzia C, Kobayashi K, Barrandon Y (2001) Morphogenesis and renewal of hair follicles from adult multipotent stem cells. Cell 104:233–245
10. Senoo M, Pinto F, Crum CP, McKeon F (2007) p63 is essential for the proliferative potential of stem cells in stratified epithelia. Cell 129:523–536
11. Shipley GD, Pittelkow MR, Wille JJ Jr, Scott RE, Moses HL (1986) Reversible inhibition of normal human prokeratinocyte proliferation by type beta transforming growth factor-growth inhibitor in serum-free medium. Cancer Res 46:2068–2071
12. Schmierer B, Hill CS (2007) TGF-β-SMAD signal transduction: molecular specificity and functional flexibility. Nat Rev Mol Cell Biol 8:970–982
13. Ikushima H, Miyazono K (2010) TGF-β signalling: a complex web in cancer progression. Nat Rev Cancer 10:415–424
14. Ghahary A, Marcoux Y, Karimi-Busheri F, Tredget EE (2001) Keratinocyte differentiation inversely regulates the expression of involucrin and transforming growth factor β1. J Cell Biochem 83:239–248
15. Suzuki D, Senoo M (2015) Dact1 regulates the ability of 3T3-J2 cells to support proliferation of human epidermal keratinocytes. J Invest Dermatol 135:2894–2897
16. Robertson IB, Rifkin DB (2013) Unchaining the beast; insights from structural and evolutionary studies on TGF-β secretion, sequestration, and activation. Cytokine Growth Factor Rev 24:355–372
17. Suzuki D, Pinto F, Senoo M (2017) Inhibition of TGF-β signaling promotes expansion of human epidermal keratinocytes in feeder cell co-culture. Wound Repair Regen 25(3):526–531. https://doi.org/10.1111/wrr.12541
18. Bullock AJ, Higham MC, MacNeil S (2006) Use of human fibroblasts in the development of a xenobiotic-free culture and delivery system for human keratinocytes. Tissue Eng 12:245–255
19. Sugiyama H, Maeda K, Yamato M, Hayashi R, Soma T, Hayashida Y, Yang J, Shirakabe M, Matsuyama A, Kikuchi A, Sawa Y, Okano T, Tano Y, Nishida K (2008) Human adipose tissue-derived mesenchymal stem cells as a novel feeder layer for epithelial cells. J Tissue Eng Regen Med 2:445–449
20. Skurk T, Ecklebe S, Hauner H (2007) A novel technique to propagate primary human preadipocytes without loss of differentiation capacity. Obesity 15:2925–2931
21. Freedberg IM, Tomic-Canic M, Komine M, Blumenberg M (2001) Keratins and the keratinocyte activation cycle. J Invest Dermatol 116:633–640
22. Rasmussen C, Thomas-Virnig C, Allen-Hoffmann BL (2013) Classical human epidermal keratinocyte cell culture. Methods Mol Biol 945:161–175
23. Lim X, Tan SH, Koh WL, Chau RM, Yan KS, Kuo CJ, van Amerongen R, Klein AM, Nusse R (2013) Interfollicular epidermal stem cells self-renew via autocrine Wnt signaling. Science 342:1226–1230

Chapter 3

Isolation and Culture of Epidermal Melanocytes

Muriel Cario and Alain Taieb

Abstract

Melanocytes which represent around 5% of epidermal cells are located in the basal layer. To culture melanocytes we used trypsin digestion instead of dispase to obtain a cell suspension containing only basal keratinocytes and melanocytes. Melanocytes are cells which need a great attention. Indeed they dedifferentiate easily in culture as soon as they are in pure culture. Factors secreted by contaminating keratinocytes allow melanocytes to stay dendritic but by regulating their number avoid their growth. In order to age, phototype and other individual dependent factors regulate the behavior of melanocytes in vitro. Thus, microscopic examination of melanocytes has to be performed each day to adapt conditions of culture to each primary cell culture. This is the secret to have a nice melanocyte culture.

Key words Melanocytes, Epidermis, Isolation, Culture, Skin

1 Introduction

During embryonic development in mammals, the neural crest-derived melanoblasts migrate along the dorsolateral axis, and cross the basal membrane separating the dermis from the epidermis to reach their final location in the interfollicular epidermis and epidermal hair follicles [1]. In interfollicular epidermis, melanocytes are located in the basal layer (Fig. 1) and represent around 10% of epidermal basal cells [2] whereas melanocyte stem cells are located in the hair follicle bulge-subbulge area in mammals [3, 4]. The density of melanocytes in unexposed lower back skin is similar whatever the skin type (Asian, black, white) and ranges from 12.2 to 12.8 melanocytes by mm [5]. However, according to skin localization, density of melanocytes differs. There are approximately 2000 melanocytes per square millimeter in the skin of the scrotum or in the foreskin and 1000–1500 melanocytes per square millimeter on the rest of non-palmoplantar skin [6]. Otherwise, palmoplantar skin contains less melanocytes than non palmoplantar skin [7, 8]. Melanocytes are dendritic cells which produce and transfer melanin via their dendrites to around 36 surrounding

Sophie Böttcher-Haberzeth and Thomas Biedermann (eds.), *Skin Tissue Engineering: Methods and Protocols*, Methods in Molecular Biology, vol. 1993, https://doi.org/10.1007/978-1-4939-9473-1_3,

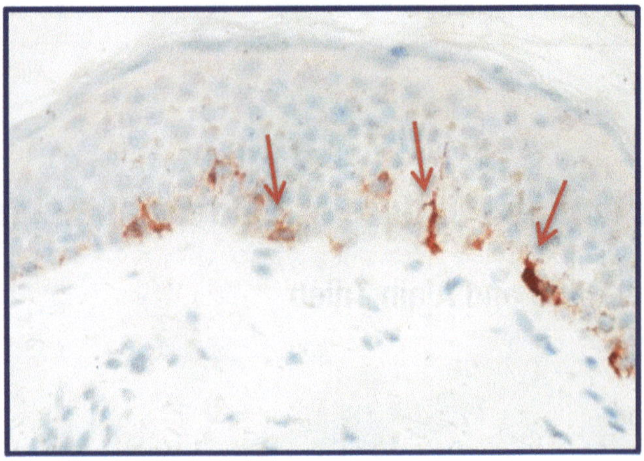

Fig. 1 Melanocyte location in human epidermis. Immunohistochemisry of human skin using anti-melanA antibody and envision system (Dako) allowed visualization of melanocytes in red (arrow). Counterstaining of nucleus was done with hematoxylin (violet)

keratinocytes (Fig. 1) [9]. This complex is named epidermal melanin unit. Melanocytes produce pheomelanin (orange melanin) and eumelanins (brown, black melanins) which concentration depends on the phototype [10–12]. Melanin production is under the control of specific enzyme tyrosinase which is common for the two melanogenic pathways whereas TRP-1 and TRP-2 are specific of the eumelanogenic pathway. Under stimulation by sunlight, melanocytes from darker skin have a greater capacity to produce melanin [13].

2 Materials

For human skin cell culture, all melanocytes are made in a biosafety level 2 culture laboratory, and all solutions are sterilized by heating or 0.22 μm filtration.

2.1 Human Skin Samples

Human skin samples are surgical wastes obtained after informed consent of donor. Common surgical wastes are mammary skin, foreskin, and abdominal skin. Samples from patients are often biopsy of 4 mm diameter.

2.2 Cell Culture

1. Hanks' balanced salt solution (HBSS).
2. Trypsin solution: 0.25% Trypsin, 0.1% EDTA in HBSS (*see* **Note 1**).
3. FBS solution: 10% Fetal bovine serum (FBS) diluted in HBSS.
4. Phosphate-buffered saline (PBS): 0.137 M NaCl, 0.05 M NaH_2PO_4, in deionized H_2O, adjust to pH 7.4.

5. Melanocyte medium: (a) Commercial MCDB153 medium: supplemented with 20 μg/mL insulin, 140 μg/mL bovine pituitary extract, 1.75 μM hydrocortisone, 3% FBS, and 1% penicillin-streptomycin. (b) Commercial M254 medium: supplemented with 3 mg/mL heparin, 10 ng/mL PMA, 3 ng/mL FGF-2, 5 mg/mL insulin, 0.2% bovine pituitary extract, 0.18 mg/mL hydrocortisone, and 1% penicillin-streptomycin. Some authors also add transferrin and FBS. (c) Commercial melanocyte growth medium M2 or melanocyte basal medium (MBM): supplemented with corresponding commercial supplement mix (C42300 + C39420 or C24210 + C39410, PromoCell) and with 1% penicillin-streptomycin. (d) Commercial MBM: could also be supplemented with animal-free 0.1 mg/mL GM-CSF, animal-free 0.1 mg/mL SCF, 0.2 mg/mL animal-free FGF-2, and 1% penicillin-streptomycin to have clinical grade medium (*see* **Notes 2** and **3**).
6. Melanocytes are isolated from skin sample (*see* Subheading 3.1). Alternatively, one can purchase commercially available melanocytes (e.g., human skin-derived melanocytes, PromoCell, ATCC).
7. 0.4% Trypan blue solution in PBS pH 7.4.
8. 3.7% Paraformaldehyde in PBS pH 7.4.
9. 40 μm Filter strainer.
10. Cell counter.
11. Chamber slide or cell culture wells (96-well plate).
12. Horseradish peroxidase (HRP).
13. 3% Hydrogen peroxide in PBS pH 7.4.
14. 4′,6-Diamidino-2-phenylindole (DAPI).

3 Methods

3.1 Isolation of Epidermal Melanocytes

1. Degrease, thin the samples, and cut into fragments with a maximum size of 0.5 × 0.5 cm (*see* **Note 4**) (Fig. 2).
2. Incubate in trypsin solution for around 3 h at 37 °C or overnight at 4 °C (*see* **Notes 5–7**).
3. Neutralize trypsin with same volume of FBS.
4. Remove the epidermis and scrape the basal layer not too hard to avoid fibroblasts in HBSS.
5. Filter the obtained suspension (40 μm filter strainer) and count cell.
6. Centrifuge at 250 × *g* for 5 min.

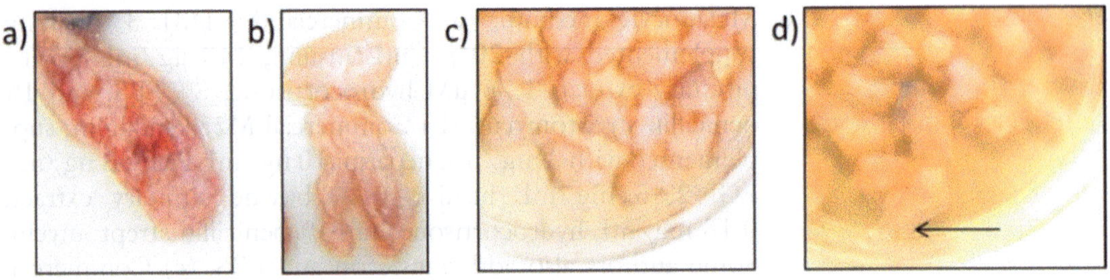

Fig. 2 Skin sample preparation. Appearance of skin before (**a**) and after (**b**) degreasing and thinning of skin. (**c**) Skin small pieces' epidermal size up in trypsin-EDTA. (**d**) Epidermis peeling off (arrow)

Fig. 3 Primo-culture of melanocyte. According to medium use more or less keratinocytes (arrow) can contaminate the culture inducing difficulty for purification of melanocytes

7. Dilute 50 μL of solution containing cells in 50 μL trypan blue and count non-blue cells.
8. Suspend the pellet in melanocyte medium.
9. Seed cells at a density of 1×10^5 per cm^2 (*see* **Notes 8** and **9**).
10. Change medium three times a week (*see* **Note 10**) after rinsing with HBSS.
11. After about a week, the primo-culture will be at confluence and melanocytes may be purified (*see* **Note 8**) and amplified (*see* **Notes 11** and **12**) (Fig. 3).

3.2 Isolation of Epidermal Melanocytes from a Punch Biopsy

1. Thin biopsy by removing hypodermis and reticular dermis and cut biopsy into two.
2. Incubate in trypsin solution for around 1–2 h at 37 °C.
3. Neutralize trypsin with same volume of FBS.
4. Remove the epidermis and scrape the basal layer in melanocyte medium.

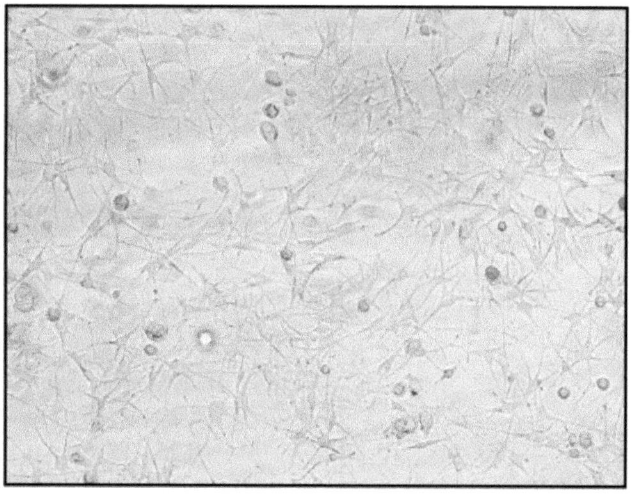

Fig. 4 Confluent low-phototype melanocytes

5. Put cell suspension in a well of a 24-well plate.
6. Change medium without rinsing 24–48 h after seeding.
7. Change medium twice a week until confluence (*see* **Note 13**) (Fig. 4).

3.3 Culture and Amplification of Melanocytes (See Notes 14 and 15)

1. Rinse primo-culture of melanocytes with HBSS to improve trypsin action when culture reaches confluence or when too much keratinocytes contaminate the culture.
2. Incubate melanocytes in the trypsin working solution (0.025% trypsin, 0.01% EDTA) less than 1 min. Follow detachment under microscope (*see* **Notes 16** and **17**).
3. Neutralize trypsin with equal volume of 10% FBS (1.5 mL for 25 cm^2).
4. Collect the solution containing cells (note the volume) and before centrifugation take 50 μL to count cells.
5. Centrifuge the solution at 250 × *g* at room temperature for 5 min.
6. Dilute 50 μL of solution containing cells in 50 μL trypan blue and count non-blue cells.
7. Seed cells in culture medium usually by doubling the surface (*see* **Note 18**).
8. Repeat **steps 1**–**7** until having enough melanocytes to perform desired study (*see* **Note 20**) (Figs. 5 and 6).

3.4 Transduction of Melanocytes

Melanocytes from patient are often difficult to obtain and thus you can model disease using gene transfer. Transduction efficiency either for overexpression or silencing lentiviral vectors is high in melanocyte.

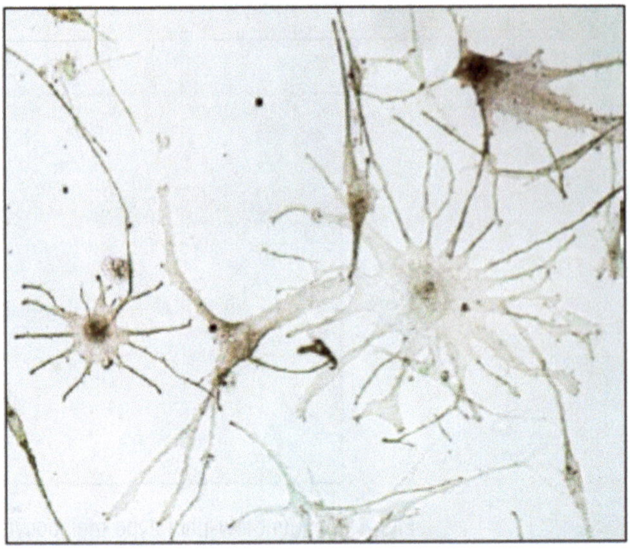

Fig. 5 Observation of melanocytes from phototype VI skin without any staining. Two populations of high dendritic melanocytes can be distinguished according to the level of intrinsic pigmentation in a culture of melanocyte of high phototype

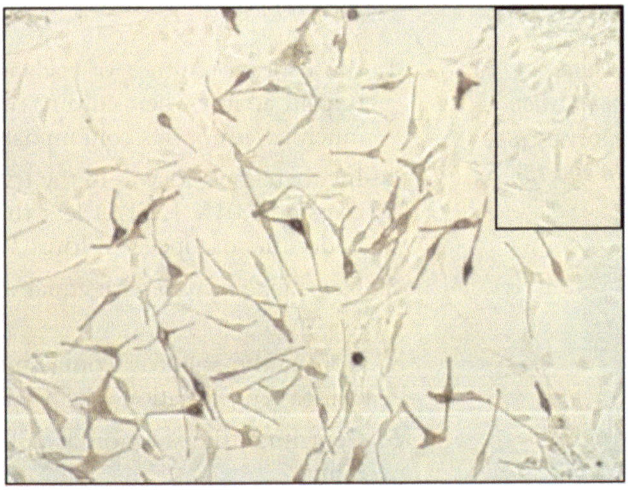

Fig. 6 Observation of culture of melanocytes and melanoblasts without any staining. Two populations can be distinguishing, according to the level of intrinsic pigmentation, shape, and organization. Melanocytes are pigmented bi- or tripolar cells whereas melanoblasts (rectangle) are unpigmented bipolar cells. Melanoblasts are more in close contact than melanocytes

1. Rinse pure culture of melanocyte passages 1–2 with HBSS to improve trypsin action.
2. Incubate melanocytes in the trypsin working solution (0.025% trypsin, 0.01% EDTA) less than 1 min. Follow detachment under microscope.

3. Neutralize trypsin with equal volume of 10% FBS (1.5 mL for 25 cm^2).
4. Collect the solution containing cells (note the volume) and before centrifugation take 50 μL to count cells.
5. Centrifuge the solution at 250 × *g* at room temperature for 5 min.
6. Dilute 50 μL of solution containing cells in 50 μL trypan blue and count non-blue cells.
7. Seed melanocytes at a density of 18,000–20,000 by cm^2.
8. 6–18 h after plating add virus at a multiplicity of infection (MOI) of 1–10 to determine the efficient MOI in a small volume of medium without FBS and incubate at 37 °C, 5% CO_2 (*see* **Note 21**).
9. 6–18 h after infection add complete melanocyte medium.
10. 48 h after infection rinse two times the culture and add complete medium.
11. Transduction is efficient 72 h after infection.
12. 100% Transduction can be achieved by fluorescent cell sorting if lentivector contains fluorescent protein, or by incubation with antibiotics (such as 10 μg/mL puromycin) if lentivector contains an antibiotic-resistant gene (*see* **Note 22**) (Fig. 7).

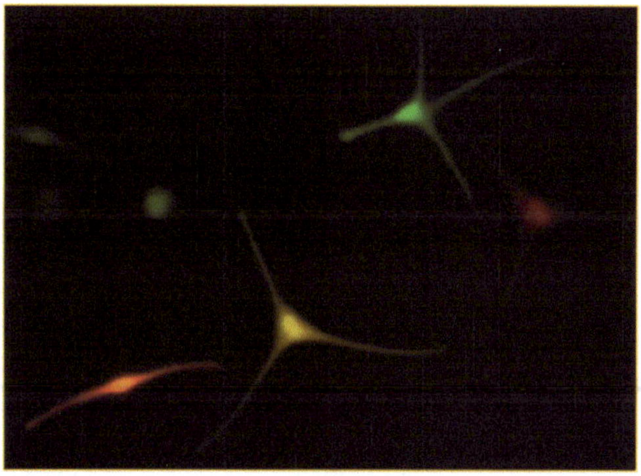

Fig. 7 Aspect of culture of melanocyte after double-transduction MOI 4. Melanocytes were transduced with a lentivector containing gene of interest Y and GFP coding sequence and a second lentivector containing another gene of interest Y' and tdTomato coding sequence. Each transduction was at MOI 2. Double-transduced melanocytes appear in yellow and single-transduced melanocytes appear in red (TdTomato) or green (GFP)

3.5 Characterization of Melanocytes by Immunohistochemistry

1. Culture cells in chamber slide or cell culture wells (96-well plate) for 6–24 h.
2. Rinse cells three times with cold PBS pH 7.4.
3. Fix cells with 3.7% paraformaldehyde for 10 min at 4 °C.
4. Rinse cells three times with cold PBS pH 7.4.
5. Permeabilize cells with 0.2% triton x-100 in PBS pH 7.4 (*see* **Note 23**).
6. Wash three times with PBS pH 7.4.
7. If horseradish peroxidase (HRP) staining is used, block endogenous peroxidase with 3% hydrogen peroxide during 10 min and wash three times with PBS pH 7.4.
8. Incubate with 2.5–10% FBS for 20 min at 4 °C (*see* **Note 24**).
9. Incubate with primary antibody (Table 1) diluted according to the manufacturer in PBS pH 7.4 for 1 h at room temperature to overnight 4 °C in a moisture chamber.
10. Collect diluted antibody and keep this dilution at the antibody storage temperature (*see* **Note 25**).
11. Wash with PBS pH 7.4 three times.
12. Incubate with secondary fluorescent antibody raised against the host species used to generate the primary antibody diluted in PBS pH 7.4 according to the manufacturer for 1 h at room temperature in the dark or with HRP-coupled secondary antibody.
13. Collect diluted antibody and keep this dilution at the antibody storage temperature (*see* **Note 25**).
14. Wash with PBS pH 7.4 three times.

Table 1
List of useful antibodies to characterize melanocyte

Antibody	Reference	Dilution
HMB45	Dako M0634	1/20°
C-Kit (CD117)	Dako A4502	1/1000°
NKI/beteb	Monosan Mon7006	1/20°
TRP-1	Abcam ab83774	1/100°
MITF	Abcam ab20663	1/100°
TRP-2	Abcam ab74073	1/500°
Tyrosinase	Novusbio NBP1	1/100°
MelanA (Mart-1)	Dako M7196	1/100°

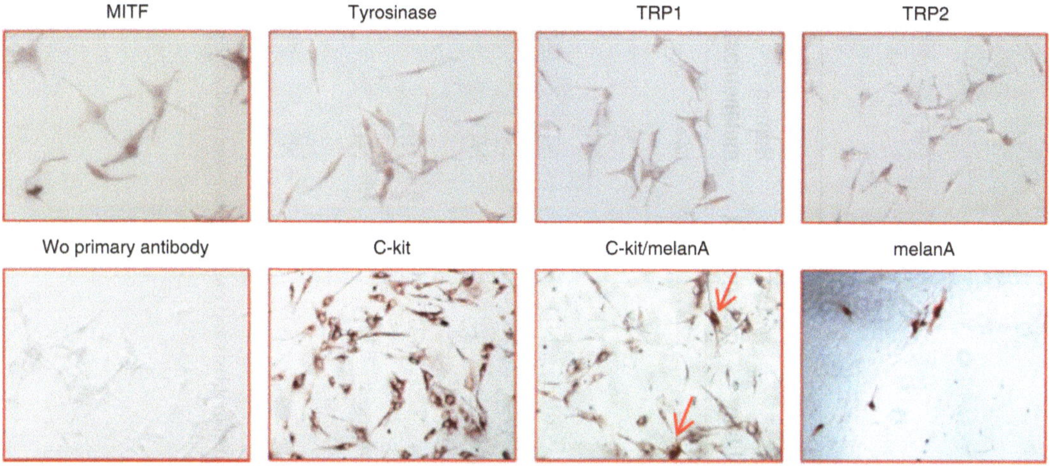

Fig. 8 Expression of various markers in melanocyte-derived cells. Except for c-Kit which chromogen was vector VIP (violet), chromogen was AEC (rouge). In the double-staining c-Kit/melanA, melanA-positive cells were identified with an arrow. MelanA staining revealed melanocytes

15. Incubate with chromogen according to the manufacturer's instructions for HRP staining.
16. Incubate with 4′,6-diamidino-2-phenylindole (DAPI) 10 min for fluorescent staining.
17. Wash with PBS pH 7.4 three times.
18. Mount with adapted mounting medium (Fig. 8, Table 2).

3.6 Evaluation of Tyrosinase Activity in Melanocytes

Melanoblasts are dopa negative; thus this technique allows differentiating melanoblasts from melanocytes by detection of tyrosinase activity [14–16].

1. Culture cells in Lab-Tek® chamber slide or cell culture wells (96-well plate) for 6–24 h.
2. Rinse cells three times with cold PBS pH 7.4.
3. Fix cells with 3.7% paraformaldehyde for 15 min at 4 °C.
4. Rinse cells three times with cold PBS pH 7.4.
5. Incubate with L-DOPA (0.1% in PBS pH 7.4) for 1 h at 37 °C.
6. Renew L-DOPA solution and incubate for 3 h at 37 °C.
7. Wash three times with PBS pH 7.4.
8. Mount in aqueous mounting medium (Fig. 9).

Table 2
Reactivity of melanocytes and melanoblasts against antibodies

Marker	Melanocyte/melanoblast (dedifferentiated melanocyte)	Melanocyte/ melanoblast	Melanocyte/ dedifferentiated melanocyte	Melanocyte	Mouse melanoblast	Mouse melanoblast	Human melanoblasts
References	[16]	[14]	[17]	[19, 20]	[16, 21]	[22]	[23]
HMB45	+/−	+/+	ND	ND	ND	ND	−
c-Kit (CD117)	−/+[a]	ND	+[a]/−	+	ND	ND	ND
TRP-1	ND	+/+	+/−	+	−/+ (according to medium	ND	ND
TRP-2	+/+	ND	ND	ND	−/+ (according to medium)	+	ND
Tyrosinase	ND	+/+[b]	ND	+	−	−[a]	ND
MelanA (Mart-1)	ND	+/+[a]	ND	ND	ND	ND	ND
S-100	+/−	ND	ND	ND	ND	ND	ND

[a]Mixed cells of positive and negative
[b]Detection of tyrosinase by RT-PCR or immunocytochemistry does not predict its activity. Indeed a great difference between melanoblasts and melanocytes is the localization of markers [14]

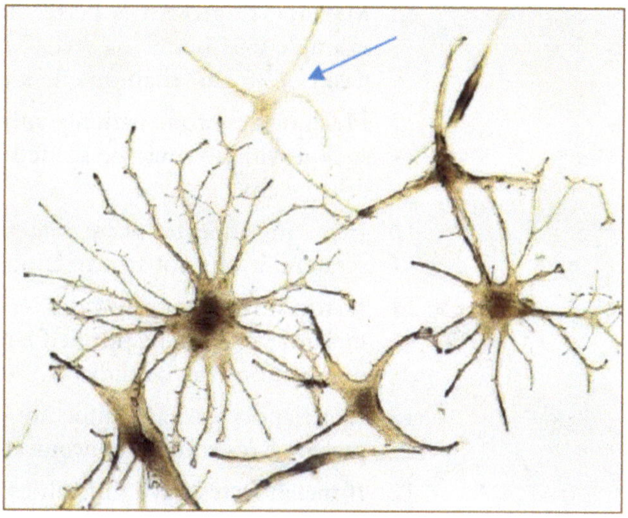

Fig. 9 Appearance of melanocytes after Dopa staining. According to Hirobe, melanobasts are Dopa-negative cells which are not to be restricted to bipolar and tripolar cells [15]. The dopa-negative cell with five dendrites may be a melanoblast

4 Notes

1. Make sure to be at 37 °C and no more to avoid inactivation of trypsin.
2. Melanin produced by high-phototype melanocytes hampers cell growth. Thus for high-phototype melanocyte M2 medium is more adapted or MCDB153 has to be supplemented with 300 μM PTU (phenyl thiourea), an inhibitor of tyrosinase. We commonly use MCDB153 without PTU for phototype I–IV melanocytes and M2 medium for phototypes V and VI.
3. Phorbol 12-myristate 13-acetate (PMA) is a skin tumor promoter, present in the supplement kit for MBM. According to experiment, it could be better to use PMA-free medium.
4. After degreasing and thinning, skin must almost seem transparent.
5. Pieces of skin must be placed in trypsin-EDTA with epidermal size up.
6. Trypsin cuts above basal layer at the difference of dispase which cuts under the basal layer at the dermo-epidermal junction.
7. Epidermis must peel off by itself. If not incubation has to be continued. This may due to the thickness or the size of the samples.

8. Melanocyte growth is better in 25 cm^2 than 75 cm^2, so for example it is better to seed 7.5 million melanocytes in three flasks of 25 cm^2 than one flask of 75 cm^2.
9. Melanocytes from patients suffering from pigment disorders such as vitiligo must be seeded at a greater density of 2×10^5 cells by cm^2.
10. If no melanocytes were seen after 2–3 days of culture, the scratching was not strong enough,
11. If too many keratinocytes contaminate the primo-culture, melanocyte must be purified before 70% confluence; otherwise purification may be difficult.
12. If melanocytes have difficulty to grow instead of amplifying melanocytes you can concentrate them in a flask of smaller size.
13. If melanocytes have difficulties to maintain in culture and are for experiments such as genetic analysis (sequencing) PMA could be added in the medium.
14. Dendricity of melanocytes varies according to phototype. Phototypes I–III are less dendritic than phototypes IV–VI.
15. Melanocytes from phototype VI have often two populations of melanocytes, one pigmented and the other unpigmented.
16. If primo-cultures are not pure, you must follow carefully cells' detachment under microscope, to avoid keratinocyte contamination.
17. If melanocytes detach too quickly and if it is difficult to purify culture, use a less concentrated solution of trypsin-EDTA as 0.0125% trypsin and 0.005% EDTA.
18. If melanocytes grow fast you have to increase the size of seeding to avoid more than one passage by week.
19. Along culture melanocytes become bigger and less dendritic. However the behavior of melanocytes in culture is individual dependent.
20. Shape of melanocytes has to be managed carefully since in culture, especially without PMA, melanocytes dedifferentiate quickly in melanoblasts [16, 17].
21. A MOI range has to be tested prior to doing the experiment since modifying gene is not harmless and may induce death or detachment of melanocytes due to genotoxicity. Indeed increasing MOI is associated with increase in vector copy number (VCN). It is admitted that there is only one copy of gene by cells when the percentage of transduction is around 30% [18].
22. Sorting or treatment with antibiotics may not be done too early since in the beginning transgene expression is due to pseudo-transduction (circular DNA) and not integrated gene.

23. Permeabilization step can be avoided if the protein is membranous and antibody targets extracellular domain.
24. Manufacturer recommends using serum from species of the secondary antibody but FBS is efficient for most staining. Incubation with 2.5–10% FBS avoids a specific fixation; the percentage has to be adjusted according to the background. Start with 2.5%. Otherwise commercial blocking solutions are available.
25. Some diluted antibodies can be used up to three times.

References

1. Bonaventure J, Domingues MJ, Larue L (2013) Cellular and molecular mechanisms controlling the migration of melanocytes and melanoma cells. Pigment Cell Melanoma Res 26(3):316–325
2. Haass NK, Herlyn M (2005) Normal human melanocyte homeostasis as a paradigm for understanding melanoma. J Investig Dermatol Symp Proc 10(2):153–163
3. Jimbow K, Quevedo WC, Fitzpatrick TB, Szabo G (1976) Some aspects of melanin biology: 1950–1975. J Invest Dermatol 67 (1):72–89
4. Nishimura EK (2011) Melanocyte stem cells: a melanocyte reservoir in hair follicles for hair and skin pigmentation. Pigment Cell Melanoma Res 24(3):401–410
5. Tadokoro T, Yamaguchi Y, Batzer J, Coelho SG, Zmudzka BZ, Miller SA et al (2005) Mechanisms of skin tanning in different racial/ethnic groups in response to ultraviolet radiation. J Invest Dermatol 124 (6):1326–1332
6. Démarchez M. Melanocyte and pigmentation. 2011. https://biologiedelapeau.fr/spip.php?article7
7. Szabo G (1967) The regional anatomy of the human integument with special reference to the distribution of hair follicles, sweat glands and melanocytes. Philos Trans R Soc Lond Ser B Biol Sci 252(779):447–485
8. Yamaguchi Y, Itami S, Watabe H, Yasumoto K-I, Abdel-Malek ZA, Kubo T et al (2004) Mesenchymal-epithelial interactions in the skin: increased expression of dickkopf1 by palmoplantar fibroblasts inhibits melanocyte growth and differentiation. J Cell Biol 165 (2):275–285
9. Fitzpatrick TB, Breathnach AS (1963) The epidermal melanin unit system. Dermatol Wochenschr 147:481–489
10. Hennessy A, Oh C, Diffey B, Wakamatsu K, Ito S, Rees J (2005) Eumelanin and pheomelanin concentrations in human epidermis before and after UVB irradiation. Pigment Cell Res 18(3):220–223
11. Hunt G, Kyne S, Ito S, Wakamatsu K, Todd C, Thody A (1995) Eumelanin and phaeomelanin contents of human epidermis and cultured melanocytes. Pigment Cell Res 8(4):202–208
12. Tobin D, Quinn AG, Ito S, Thody AJ (1994) The presence of tyrosinase and related proteins in human epidermis and their relationship to melanin type. Pigment Cell Res 7(4):204–209
13. Quevedo WC, Fitzpatrick TB, Pathak MA, Jimbow K (1975) Role of light in human skin color variation. Am J Phys Anthropol 43 (3):393–408
14. Cook AL, Donatien PD, Smith AG, Murphy M, Jones MK, Herlyn M et al (2003) Human melanoblasts in culture: expression of BRN2 and synergistic regulation by fibroblast growth factor-2, stem cell factor, and endothelin-3. J Invest Dermatol 121 (5):1150–1159
15. Hirobe T (1992) Melanocyte stimulating hormone induces the differentiation of mouse epidermal melanocytes in serum-free culture. J Cell Physiol 152(2):337–345
16. Zhao Z, Jin C, Ding K, Ge X, Dai L (2012) Dedifferentiation of human epidermal melanocytes into melanoblasts in vitro. Exp Dermatol 21(7):504–508
17. Kormos B, Belső N, Bebes A, Szabad G, Bacsa S, Széll M et al (2011) In vitro dedifferentiation of melanocytes from adult epidermis. PLoS One 6(2):e17197
18. Kustikova OS, Wahlers A, Kühlcke K, Stähle B, Zander AR, Baum C et al (2003) Dose finding with retroviral vectors: correlation of retroviral vector copy numbers in single cells with gene transfer efficiency in a cell population. Blood 102(12):3934–3937

19. Hachiya A, Kobayashi A, Ohuchi A, Takema Y, Imokawa G (2001) The paracrine role of stem cell factor/c-kit signaling in the activation of human melanocytes in ultraviolet-B-induced pigmentation. J Invest Dermatol 116 (4):578–586
20. Luo D, Chen H, Searles G, Jimbow K (1995) Coordinated mRNA expression of c-Kit with tyrosinase and TRP-1 in melanin pigmentation of normal and malignant human melanocytes and transient activation of tyrosinase by Kit/SCF-R. Melanoma Res 5(5):303–309
21. Hirobe T, Osawa M, Nishikawa S-I (2004) Hepatocyte growth factor controls the proliferation of cultured epidermal melanoblasts and melanocytes from newborn mice. Pigment Cell Res 17(1):51–61
22. Sviderskaya EV, Wakeling WF, Bennett DC (1995) A cloned, immortal line of murine melanoblasts inducible to differentiate to melanocytes. Development 121(5):1547–1557
23. Holbrook KA, Underwood RA, Vogel AM, Gown AM, Kimball H (1989) The appearance, density and distribution of melanocytes in human embryonic and fetal skin revealed by the anti-melanoma monoclonal antibody, HMB-45. Anat Embryol (Berl) 180 (5):443–455

Chapter 4

Long-Term Expansion of Mouse Primary Epidermal Keratinocytes Using a TGF-β Signaling Inhibitor

Filipa Pinto, Daisuke Suzuki, and Makoto Senoo

Abstract

Mouse models have been used to study the physiology and pathogenesis of the skin. However, propagation of mouse primary epidermal keratinocytes remains challenging. In this chapter, we introduce a newly developed protocol that enables long-term expansion of $p63^+$ mouse epidermal keratinocytes in low-Ca^{2+} media without the need of progenitor cell purification steps or support by a feeder cell layer. Pharmacological inhibition of TGF-β signaling in crude preparations of mouse epidermis robustly increases proliferative capacity of $p63^+$ epidermal progenitor cells while preserving their ability to differentiate. Suppression of TGF-β signaling also permits $p63^+$ epidermal keratinocytes to form macroscopically large clones in 3T3-J2 feeder cell co-culture. This simple and efficient approach will facilitate the use of mouse models by providing $p63^+$ primary epidermal keratinocytes in quantity.

Key words Mouse models, Primary epidermal keratinocytes, Transcription factor p63, TGF-signaling, Small-molecule inhibitors

1 Introduction

Mouse models have been used in studies of normal and disease conditions of the skin. However, the growth of primary epidermal keratinocytes derived from mice rapidly declines in culture and the cells become terminally differentiated [1], a limit that restricts the use of mouse primary keratinocytes for functional analyses of the skin. As proliferation and differentiation of epithelial cells are tightly coupled through the induction of cyclin-dependent kinase (CDK) inhibitor genes [2–4], suppression of growth arrest and of premature differentiation are both potential approaches to improve the life span of mouse primary epidermal keratinocytes [2, 5–8].

Although the use of low-calcium (Ca^{2+}) media has extended proliferation of mouse epidermal keratinocytes in the short term [1], the most effective protocols to propagate these primary cells rely on modified 3T3-J2 feeder co-culture [5, 7] or use of

Sophie Böttcher-Haberzeth and Thomas Biedermann (eds.), *Skin Tissue Engineering: Methods and Protocols*, Methods in Molecular Biology, vol. 1993, https://doi.org/10.1007/978-1-4939-9473-1_4,

fluorescence-activated cell sorting (FACS)-purified progenitor populations with triple-drug inhibitors [8]. Highly efficient protocols that eliminate the use of undefined factors (e.g., feeder cells), labor-intensive purification steps, and potentially complex effects of multiple drugs will facilitate the use of primary cells of mice in studies of skin biology.

Transforming growth factor-β (TGF-β) signaling regulates proliferation and differentiation of many different epithelial progenitor cells [9], including those that are controlled by the transcription factor p63 (e.g., epidermal keratinocytes) [10]. TGF-β signaling is mediated through the receptor complex consisting of the type I TGF-β receptor (TGFβR1/ALK5) and the type 2 TGF-β receptor (TGFβR2) [11]. Upon binding of the TGF-β ligands, TGFβR2 phosphorylates and activates TGFβR1/ALK5, resulting in the phosphorylation and nuclear translocation of Smad2/3, downstream effectors of TGF-β signaling [12]. We have shown recently that inhibition of TGF-β signaling suppresses Smad2/3 nuclear localization and consequently increases the proliferative capacity of $p63^+$ mouse epidermal keratinocytes in vitro [13].

In this chapter, we introduce a newly developed protocol that enables long-term proliferation of $p63^+$ mouse primary epidermal keratinocytes utilizing a TGF-β signaling inhibitor without the need of a feeder layer or progenitor cell purification steps. Suppression of TGF-β signaling also permits the expanded $p63^+$ mouse epidermal keratinocytes to form macroscopically large clones in 3T3-J2 feeder cell co-culture for clonal evaluation. We anticipate that this simple and efficient approach will facilitate the use of mouse models for studying the physiology and pathogenesis of the epidermis.

2 Materials

2.1 Biological and Other Materials

1. C57BL/6 mice (Jackson Laboratories and Charles River Laboratories) were used to develop the protocol introduced in this chapter. Animal work described below requires an approval by local Institutional Animal Care and Use Committees (IACUC).
2. 3T3-J2 cells [14] were provided by Howard Green at Harvard Medical School.
3. Isojin (Meiji Seika Pharma Co., Ltd.)

2.2 Instruments and Supplies

1. Humidified CO_2 incubators (5 and 10%).
2. CO_2 gas cylinders.
3. 37 °C Water bath.
4. Hemocytometer.

5. Inverted microscope.
6. γ-Irradiator with a Cesium-137 source.
7. Tissue culture plates (35, 60, 100 mm).
8. 15 and 50 mL polypropylene conical tubes.
9. 1.5 mL Eppendorf tubes.
10. Cryovials.
11. Syringe filter unit (0.2 μm).
12. 70 μm cell strainer.
13. Parafilm.
14. Whatman filter paper.

2.3 Cell Culture Media and Freezing Media

1. CnT-PR medium: 100 U/mL Penicillin, 100 μg/mL streptomycin in CnT-PR media (CnT-Prime epithelial culture medium).
2. DMEM: 4.5 g/L Glucose, 584 mg/L L-glutamine, 110 mg/mL sodium pyruvate in Dulbecco's modified Eagle's medium (DMEM).
3. J2 medium: 100 U/mL Penicillin, 100 μg/mL streptomycin, 10% (v/v) CS (bovine calf serum) in DMEM [14].
4. FAD medium: Mix DMEM and Ham's F-12 media at a 3:1 ratio (v/v) and supplement with 10% v/v fetal bovine serum (FBS), 100 U/mL penicillin, and 100 μg/mL streptomycin [14].
5. Complete FAD (cFAD) medium: 10 ng/mL EGF, 5 μg/mL insulin, 2×10^{-9} M T_3, 0.4 μg/mL hydrocortisone, 24 μg/mL adenine, 1×10^{-10} M cholera toxin in FAD medium [14].
6. Freezing medium: Mix cell culture media above (CnT-PR medium, J2 medium, or FAD medium), serum (CS or FBS), and dimethyl sulfoxide (DMSO) at a 4:5:1 ratio (v/v/v). To freeze 3T3-J2 cells use a combination of J2 medium and CS. To freeze epidermal cells, use either CnT-PR or FAD medium along with FBS.

2.4 Cell Culture Reagents

1. Adenine stock: 24 mg/mL Adenine in 0.2 N HCl. Sterilize with a 0.2 μm syringe filter. Store in 0.5 mL aliquots at −20 °C.
2. T_3 (3,3′,5-triiodo-L-thyronine sodium salt) stock: 10 μM T_3 in PBS (phosphate-buffered saline without calcium and magnesium). Sterilize with a 0.2 μm syringe filter. Store in 0.5 mL aliquots at −20 °C.
3. Hydrocortisone stock: 5 mg/mL Hydrocortisone in 95% EtOH. Store in 0.2 mL aliquots at −20 °C.

4. Cholera toxin stock: 0.1 mg/mL Cholera toxin in dH_2O. Sterilize with a 0.2 μm syringe filter. Store in 0.4 mL aliquots at −80 °C.
5. Insulin stock: 0.1 mg/mL Insulin in 0.005 N HCl. Sterilize with a 0.2 μm syringe filter. Store in 0.5 mL aliquots at −80 °C.
6. EGF stock: 20 μg/mL EGF (human epidermal growth factor) in 1% bovine serum albumin (BSA) aqueous solution. Sterilize with a 0.2 μm syringe filter. Store in 0.5 mL aliquots at −80 °C.
7. Inhibitor stocks: 25 mM Inhibitors (RepSox, A83-01, LY364947, SB525334, and SB431542) in dimethyl sulfoxide (DMSO). Store in 25 μL aliquots at −80°C. In the protocol described in this chapter, only RepSox is required.
8. 1% Rhodamine B: 1% (w/v) Rhodamine B in dH_2O, filter through Whatman filter papers, and store at room temperature (RT).

3 Methods

3.1 Collection of Crude Primary Epidermal Cells from Mouse Skin

1. Euthanize embryos to neonatal mice by decapitation. Postnatal mice should be euthanized by overdosing of isoflurane or CO_2 inhalation, followed by cervical dislocation.
2. When mice older than 2 weeks old are used, hair should be clipped prior to isolation of the skin.
3. Harvest 1–2 cm^2 back skin per animal as a single sheet of tissue using sterilized surgical scissors.
4. Sterilize the isolated skin by immersing in Isojin solution for 10 s, followed by three washes in PBS.
5. Place the skin on a sterilized 2 × 2 cm Whatman filter paper with the dermis facing down in a 35 mm tissue culture plate. The dermis is a viscous tissue and adheres tightly to the dried filter paper.
6. Slowly add 2 mL fresh 0.25% trypsin-EDTA in the tissue culture plate above, enough to soak the filter paper, and incubate for 2–4 h at RT in a tissue culture biosafety cabinet.
7. Carefully separate the epidermis from the dermis using sterilized forceps with fine tips. While the epidermis easily peels off as a whitish tissue, the dermis, as a translucent tissue, remains attached to the filter paper.
8. Transfer the isolated epidermis to 1.5 mL Eppendorf tubes.
9. Add 1 mL 0.25% trypsin-EDTA to the tubes and incubate for 5 min at RT.
10. Add 100 μL FBS (equivalent to 10% the volume of trypsin used) to inactivate the enzymatic reaction, vigorously suspend

the cells by pipetting, and spin down the cells at 400 × *g* for 5 min.

11. Resuspend the cells in 1 mL PBS and filter through a 70 μm cell strainer.
12. Centrifuge the cells at 400 × *g* for 5 min.
13. Wash the cells two additional times in PBS in order to completely remove FBS, which potentially inhibits epidermal cell growth in the subsequent culture.
14. Resuspend the cells in 100 μL CnT-PR media and count the cell number using a hemocytometer.

3.2 Enrichment and Expansion of Mouse Primary Epidermal Keratinocytes in Serum-Free Media

1. Plate 1 × 10^6 crude primary cells (*see* Subheading 3.1) in one 35 mm tissue culture plate (or one well in a 6-well plate) that contains 3 mL CnT-PR media.
2. Add RepSox to the culture media at a final concentration of 1 μg/mL (*see* **Note 1**).
3. Place the tissue culture plate in a humidified incubator with 5% CO_2 at 37 °C. Allow the cells to adhere to the plastic plate for 2–3 days (*see* **Note 2**).
4. After the incubation, replace media with fresh 3 mL CnT-PR media. Culture media should be changed every 2–3 days thereafter until the cells become subconfluent (*see* **Note 3**).
5. When the cells become subconfluent, remove culture media and wash the cells twice in 3 mL PBS.
6. Add 1 mL of 0.25% trypsin-EDTA to the cells and incubate the plate in a humidified chamber with 5% CO_2 at 37 °C for 5 min (*see* **Note 4**).
7. Tap the tissue culture plate to dislodge the cells.
8. Dissociate the cells by gentle pipetting, collect them in a 15 mL polypropylene tube, and neutralize the enzymatic reaction by adding 100 μL FBS (equivalent to 10% the volume of trypsin used).
9. Centrifuge the cells at 200 × *g* for 5 min.
10. Remove supernatant, wash the cells twice in 5 mL PBS, and resuspend in 1 mL CnT-PR media.
11. Count cell numbers. Approximately 3–5 × 10^5 cells are collected from one subconfluent 35 mm tissue culture plate.
12. Replate the cells at a 1:10 ratio in CnT-PR media containing 1 μg/mL RepSox. These secondary cultures become subconfluent in 5–7 days for subsequent experiments. By passage (P) 1 to P2, virtually all growing cells are cytokeratin $(CK)^{+}p63^{+}$ epidermal progenitor cells [13] (*see* **Note 5**).

13. To prepare frozen stock, resuspend the cell pellet in CnT-PR-based freezing media at the density of 2×10^6 cells per mL and aliquot 0.5 mL per cryovial (1×10^6 cells per vial). Place the cryovials in −80 °C overnight and transfer them to a liquid nitrogen tank for longer storage.

3.3 Expansion of Mouse Epidermal Keratinocytes in Co-culture with 3T3-J2 Feeder Cells for Clonal Analysis

3T3-J2 cells are expanded in J2 media under 10% CO_2 atmosphere. Keep in mind that the conditions of feeder cells, such as the viability, density, and passage number (*see* **Note 6**), are critical for successful expansion of mouse epidermal keratinocytes in feeder cell co-culture.

3.3.1 Culture, Passage, and Storage of 3T3-J2 Cells

1. To expand 3T3-J2 cells, one cryovial containing 1×10^6 cells is taken out of a liquid nitrogen tank and thawed in a 37°C water bath for 3 min.
2. Spin down the cells at 200 × *g* for 5 min.
3. Remove freezing media and suspend the cells in 1 mL J2 media.
4. Plate the resuspended cells in two 100 mm tissue culture plates that contain 10 mL J2 media (5×10^5 cells per plate).
5. Place tissue culture plates in a humidified incubator with 10% CO_2 at 37 °C.
6. After overnight incubation (>16 h), replace media with 10 mL of fresh J2 media. Culture media should be changed every 2–3 days thereafter until the cells become subconfluent (*see* **Note 7**).
7. When the cells become subconfluent, remove culture media and wash the cells twice in 5 mL PBS.
8. Add 2 mL 0.25% trypsin-EDTA to the cells and incubate the plates in a humidified incubator with 10% CO_2 at 37°C for 5 min.
9. Dissociate the cells by gentle pipetting, collect them in a 15 mL polypropylene tube, and neutralize the enzymatic reaction by adding the same volume (2 mL) of J2 media.
10. Centrifuge the cells at 200 × *g* for 5 min.
11. Remove supernatant and resuspend the cells in 1 mL J2 media.
12. Count cell numbers. Approximately 3×10^6 cells are collected from one subconfluent 100 mm tissue culture plate.
13. Replate 3T3-J2 cells at a 1:10–1:15 ratio at the density of 3.3×10^3 cells per cm^2 into 100 mm tissue culture plates that contain 10 mL J2 media. These secondary cultures become subconfluent in 5–7 days.
14. To prepare frozen stock, resuspend the cell pellet in freezing media at the density of 2×10^6 cells per mL and aliquot 0.5 mL

Table 1
Number of 3T3-J2 cells and the volume of cell culture media

Culture vessel	Diameter (mm)	Area (cm^2)	Volume of media (mL)	Number of 3T3-J2 cells
35 mm Plate	35	9.6	3	5.0×10^5 cells
60 mm Plate	52	21	5	1.1×10^6 cells
100 mm Plate	84	55	10	3.0×10^6 cells

per cryovial (1×10^6 cells per vial). Place the cryovials at −80 °C overnight and transfer them to a liquid nitrogen tank for longer storage.

3.3.2 Preparation of Feeder Cell Layers

Prior to seeding mouse epidermal cells, feeder cells will be mitotically arrested by either lethal γ-irradiation or mitomycin C treatment (*see* **Note 8**) to prevent propagation of the feeder cells in subsequent passages. In this protocol, we describe the use of lethally γ-irradiated feeder cells on the scale of a 60 mm tissue culture plate as an example. When other sizes of cell culture vessels are used, the number of feeder cells seeded and volume of the cell culture media used should be adjusted proportionally based on the surface area of the culture vessels (*see* **Note 9**).

1. Follow the procedures described in Subheading 3.3.1 (also Table 1) and prepare desired numbers of 60 mm tissue culture plates with subconfluent 3T3-J2 cells.
2. When 3T3-J2 cells become subconfluent, seal the culture plates with Parafilm to maintain sterility and γ-irradiate the cells at a dose of 60 Gy (6000 rads).
3. After irradiation, replace media with 5 mL of fresh J2 media and place the plates in a humidified incubator until use (*see* **Note 10**).

3.3.3 Setting Up Co-culture

1. Prepare a minimum amount of cFAD media that can be consumed in 1 week as it contains proteins and chemicals that may degrade once diluted in media.
2. Prior to seeding mouse epidermal cells, replace J2 media with 5 mL cFAD media and place the plates in a humidified incubator with 5% CO_2 at 37 °C for 1–2 h.
3. By following the steps in Subheading 3.2, prepare mouse epidermal cells in suspension. Use cFAD instead of CnT-PR media to suspend mouse epidermal cells.
4. Count viable cell numbers with the aid of trypan blue solution.
5. Adjust the density of epidermal cells to 10^3–10^5 viable cells per mL by diluting in cFAD media.

6. Seed the desired number of epidermal cells onto feeder cell layers. When rapid confluency of keratinocytes is planned, 10^3–10^4 cells per 60 mm tissue culture plate can be seeded (*see* **Note 11**). Seeding at a higher density than 10^4 cells per 60 mm plate is not recommended as it does not fully allow relatively immature epidermal keratinocytes to expand. When the determination of the proliferative capacity of epidermal cells at a clonal level is required, 10^2–10^3 cells per 60 mm plate should be seeded to permit clonal growth of individual epidermal cells (*see* **Note 12**).
7. Add 1 μM RepSox in cFAD media. Note that RepSox treatment significantly boosts the growth of mouse epidermal cells with high proliferative potential [13] (*see* **Note 13**).
8. cFAD media containing RepSox should be replaced every 2–3 days until epidermal cell clones reach desired clone sizes for evaluation.

3.3.4 Visualization of Mouse Epidermal Keratinocyte Clones in Feeder Cell Co-culture by Rhodamine B Staining

1. Wash the 60 mm co-culture plates above (Subheading 3.3.3) twice in 5 mL PBS and fix them in 10% buffered formalin for 15 min at RT.
2. Remove formalin solution and wash the cells twice in 5 mL PBS.
3. Add 5 mL of 1% Rhodamine B solution to the cells and stain for more than 15 min at RT.
4. Pour off Rhodamine B solution into a separate container (*see* **Note 14**).
5. Gently rinse the cells with tap water.
6. Invert the plates to air-dry.
7. Take photographs of the stained plates (*see* Fig. 4) and determine the sizes of individual clones as needed using NIH ImageJ or Adobe Photoshop software [13].

4 Notes

1. RepSox treatment robustly increases the CK^+ epidermal cell number in a dose-dependent manner (*see* Fig. 1). Other widely used TGF-β signaling inhibitors also stimulate the growth of CK^+ mouse primary epidermal cells in a dose-dependent manner. However, we found that RepSox treatment showed the highest increase in epidermal cell number at all the doses we examined (Fig. 1). In addition, $p63^+$ mouse epidermal progenitor cells grow at a constant rate for at least 60 days in the presence of RepSox (Fig. 2a). Proliferation of mouse epidermal cells is RepSox dependent, as removal of RepSox from the

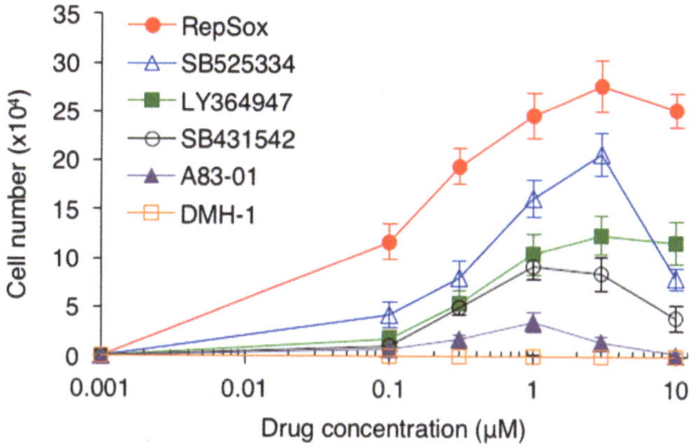

Fig. 1 Dose-dependent effects of TGF-β signaling inhibitors on cell proliferation of mouse primary epidermal keratinocytes. Newborn mouse-derived, CnT-PR-expanded cytokeratin-positive (CK^+) epidermal cells (2×10^4) were cultivated in CnT-PR media for 10 days in the presence of increasing concentration of TGF-β signaling inhibitors (RepSox, SB525334, LY364947, SB431542, and A83-01). A BMP signaling inhibitor DMH-1 was included as a negative control. Data shown are mean ± s.e.m. ($n = 3$). Adopted from reference [13]

culture results in a rapid decline in the growth of epidermal cells (Fig. 2a), accompanied by elevated expression of phosphorylated Smad2/3 at both P5 and P20 stages (Fig. 2b, c).

2. Although the majority of the cells are still floating in the culture media 3 days after the inoculation, approximately 10% of the cells adhere to the plates and initiate the growth.
3. Do not overgrow mouse epidermal cells as confluent epidermal cells are difficult to dissociate by trypsinization and they tend to differentiate prematurely in the subsequent passage. The days for primary epidermal cells to become confluent in culture seem to depend on the age of mice (*see* Fig. 3).
4. Accutase or TrypLE select can be used to more gently dissociate the cells compared with trypsin-EDTA.
5. Inhibition of TGF-β signaling also enriches and expands other $p63^+$ epithelial progenitor cells in primary crude cultures of multiple epithelia, including the cornea, oral and lingual epithelia, salivary gland, esophagus, thymus, and bladder [13].
6. We routinely expand 3T3-J2 cells at passages 4–7 and prepare a large number of frozen stocks before initiating epidermal cell co-culture. We typically use 3T3-J2 cells in co-culture at passages 6–10.
7. Do not allow 3T3-J2 cells to become confluent as the morphology of the cells is altered and the potential of feeder cells to support keratinocyte growth may decline.

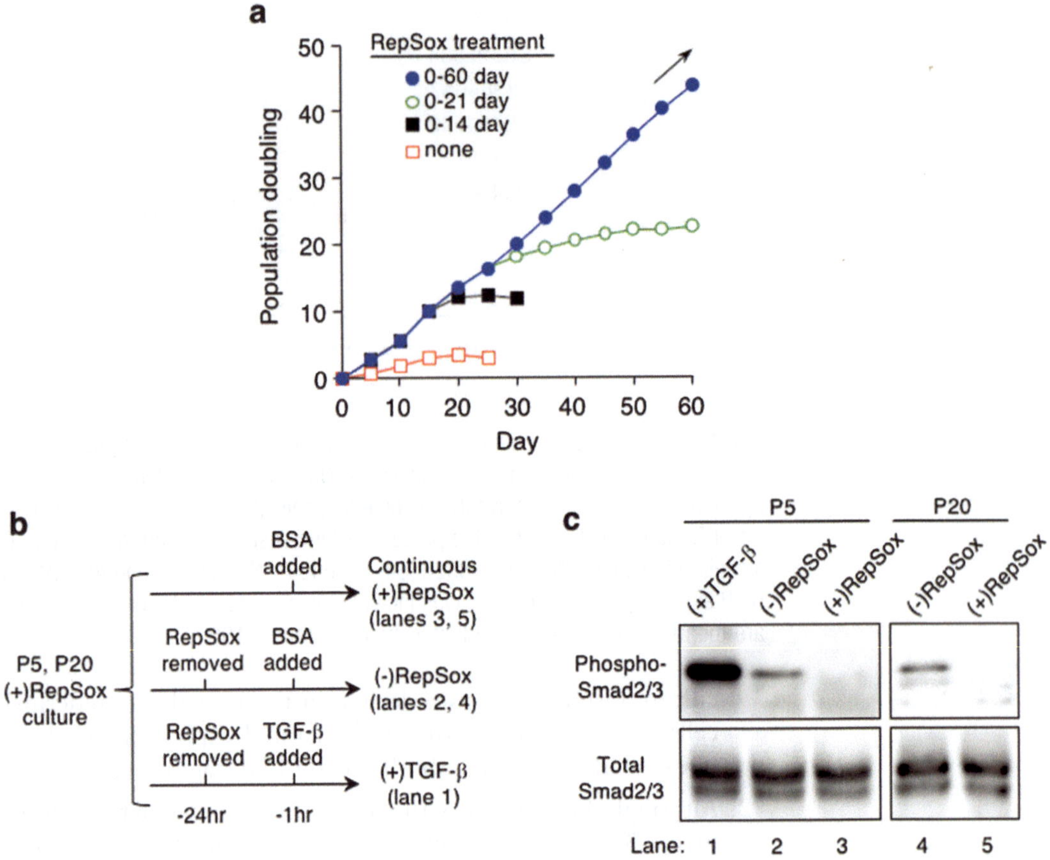

Fig. 2 TGF-β signaling inhibition-mediated enhancement of population doubling of p63$^+$ mouse primary epidermal keratinocytes. (**a**) Population doubling of newborn mouse-derived, RepSox-expanded p63$^+$ epidermal cells grown in CnT-PR media in the presence of 1 μM RepSox for 0, 14, 21, and 60 days. Data shown are representative of three independent experiments with similar results. Arrow indicates continuous cell growth. (**b**, **c**) Removal of RepSox associates with increased Smad2/3 phosphorylation. (**b**) Experimental design. Newborn mouse-derived, RepSox-expanded P5 and P20 epidermal keratinocytes were further grown in continuous presence (upper) or absence (middle and lower) of 1 μM RepSox for 24 h. Culture of P5 cells stimulated with 1 ng/mL TGF-β for 1 h prior to lysis (lower) was used as a positive control. (**c**) Expression of total and phosphorylated Smad2/3 was determined by Western blot. Lane numbers correspond to those in (**b**). Antibodies used were rabbit anti-phosphorylated Smad2/3 (D27F4, Cell Signaling Technology), rabbit anti-Smad2/3 (D7G7, Cell Signaling Technology), and horseradish peroxidase (HRP)-conjugated goat anti-rabbit IgG (Cell Signaling Technology). Adopted from reference [13]

8. Instead of γ-irradiation, mitomycin C treatment can be used to induce growth arrest in feeder cells as follows.
   (a) Add mitomycin C to 3T3-J2 media in a 50 mL tube at a final concentration of 10 μg/mL.
   (b) Aspirate cell culture media from the plates and add 3 mL of freshly prepared mitomycin C-containing feeder cell media in each 60 mm plate.

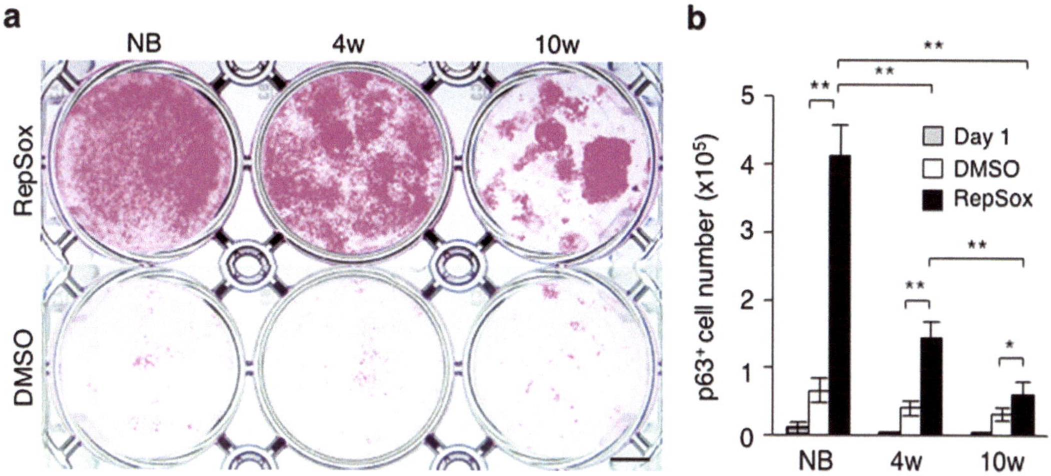

Fig. 3 Age-dependent decline in TGF-β signaling inhibition-mediated cell growth of mouse primary epidermal keratinocytes. (**a**) Representative images of Rhodamine B staining of crude culture of epidermal cells harvested from newborn (NB), 4-week-old (4w), and 10-week-old (10w) mice and grown in CnT-PR media in the presence (upper) or absence (lower) of 1 μM RepSox for 14 days. One million primary cells were seeded. Data shown are representative of two independent experiments with similar results. Bar = 5 mm. (**b**) Total number of $p63^+$ epidermal cells. One million primary cells were seeded and adherent cells were counted at day 1 in replicative wells (gray bars) as the majority of the cells remained in suspension. Solid bars and open bars represent total numbers of $p63^+$ epidermal cells at day 14 of culture in the subsequent presence of 1 μM RepSox or 0.1% DMSO, respectively ($n = 3$). $^*P < 0.05$; $^{**}P < 0.01$. Adopted from reference [13]

(c) Place the plates in a humidified incubator with 10% CO_2 at 37 °C for 2 h.

(d) Wash the cells three times in 3 mL PBS.

(e) Add 5 mL J2 media and return the plates to a humidified incubator with 10% CO_2 at 37 °C until use.

9. *See* Table 1 for expected numbers of subconfluent feeder cells at the time of epidermal cell inoculation and appropriate amounts of cell culture media used for feeder cell co-culture. When 3T3-J2 cells are seeded at the density of 3.3×10^3 cells per cm^2, it will take 5–7 days for them to become subconfluent. Higher numbers of 3 T3-J2 cells can be seeded to obtain subconfluent feeder cell layers sooner. In such a circumstance, up to 6×10^5 viable cells per 60 mm plate can be seeded and the cells can be γ-irradiated in 1–2 days when the feeder cells become subconfluent and adhere tightly to the culture vessels.
10. Although 3T3-J2 cells treated with γ-irradiation or mitomycin C can remain viable for a few weeks, we typically use them in 5–7 days after the treatment. Feeder cell media should be replaced every 2–3 days until use.
11. Epidermal cells grow faster when they are plated at higher densities, possibly due to an autocrine effect on cell proliferation [15].

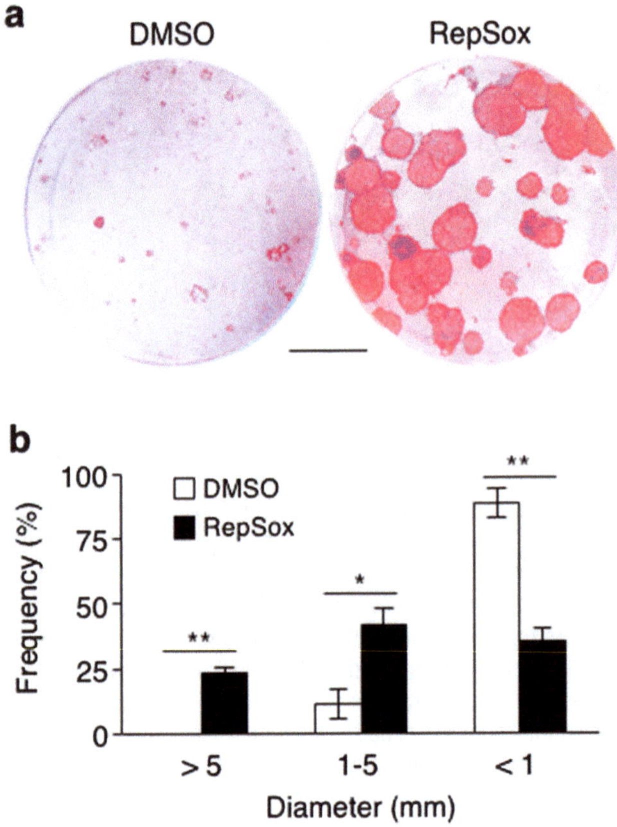

Fig. 4 Clonal expansion of mouse primary epidermal keratinocytes in co-culture with 3T3-J2 feeder cells. (**a**) Rhodamine B staining of newborn mouse-derived, RepSox-expanded P1 epidermal keratinocytes grown in 3T3-J2 co-culture in the presence or absence of 1 μM RepSox for 14 days. Bar = 10 mm. (**b**) Distribution of epidermal clone sizes at day 14. Data shown are mean ± s.e.m. ($n = 3$). *$P < 0.05$; **$P < 0.01$. Adopted from reference [13]

12. At this plating density, small colonies comprised of 4–10 cells begin to form at days 4–5. By day 7, epidermal clones become macroscopically visible and they reach 5–15 mm in diameter by day 14 in the presence of 1 μM RepSox (*see*, for example, Fig. 4). Cell culture vessels smaller than 35 mm plates (or 6-well plates) are not recommended for clonogenic analysis as epidermal cell colonies may fuse before they reach the size suitable for clonal evaluation.
13. Mouse primary epidermal cells expanded by RepSox treatment in feeder cell co-culture maintain the original high proliferative capacity as determined by serial passages in clonogenic culture with 3T3-J2 cells [13]. In addition, mouse primary epidermal cells expanded by RepSox treatment are capable of differentiation in response to Ca^{2+} stimulation [13].

14. Rhodamine B solution can be reused multiple times. Used solution should be collected in a separate container and filtered through Whatman filter papers to remove debris before use.

Acknowledgments

The authors thank Jeff Holcombe for proofreading of the manuscript. This study was supported by an R01AR066755 grant from the National Institute of Arthritis and Musculoskeletal and Skin Diseases of the National Institute of Health to M.S.

References

1. Lichti U, Anders J, Yuspa SH (2008) Isolation and short-term culture of primary keratinocytes, hair follicle populations and dermal cells from newborn mice and keratinocytes from adult mice for in vitro analysis and for grafting to immunodeficient mice. Nat Protoc 3:799–810
2. Ha L, Ponnamperuma RM, Jay S, Ricci MS, Weinberg WC (2011) Dysregulated ΔNp63α inhibits expression of Ink4a/arf, blocks senescence, and promotes malignant conversion of keratinocytes. PLoS One 6:e21877. https://doi.org/10.1371/journal.pone.0021877
3. Missero C, Di Cunto F, Kiyokawa H, Koff A, Dotto GP (1996) The absence of p21$^{Cip/WAF1}$ alters keratinocyte growth and differentiation and promotes ras-tumor progression. Genes Dev 10:3065–3075
4. Paramio JM et al (2001) The ink4a/arf tumor suppressors cooperate with p21$^{cip1/waf}$ in the processes of mouse epidermal differentiation, senescence, and carcinogenesis. J Biol Chem 276:44203–44211
5. Chapman S, McDermott DH, Shen K, Jang MK, McBride AA (2014) The effect of Rho kinase inhibition on long-term keratinocyte proliferation is rapid and conditional. Stem Cell Res Ther 5:60. https://doi.org/10.1186/scrt449
6. King KE et al (2003) ΔNp63α functions as both a positive and a negative transcriptional regulator and blocks in vitro differentiation of murine keratinocytes. Oncogene 22:3635–3644
7. Liu X et al (2012) ROCK inhibitor and feeder cells induce the conditional reprogramming of epithelial cells. Am J Pathol 180:599–607
8. Mou H et al (2016) Dual SMAD signaling inhibition enables long-term expansion of diverse epithelial basal cells. Cell Stem Cell 19:217–231
9. Watabe T, Miyazono K (2009) Roles of TGF-β family signaling in stem cell renewal and differentiation. Cell Res 19:103–115
10. Senoo M, Pinto F, Crum CP, McKeon F (2007) p63 is essential for the proliferative potential of stem cells in stratified epithelia. Cell 129:523–536
11. Schmierer B, Hill CS (2007) TGFβ-SMAD signal transduction: molecular specificity and functional flexibility. Nat Rev Mol Cell Biol 8:970–982
12. Ikushima H, Miyazono K (2010) TGFβ signalling: a complex web in cancer progression. Nat Rev Cancer 10:415–424
13. Suzuki D, Pinto F, Senoo M (2017) Inhibition of TGF-β signaling supports high proliferative potential of diverse p63^{+} mouse epithelial progenitor cells in vitro. Sci Rep 7:6089. https://doi.org/10.1038/s41598-017-06470-y
14. Rheinwald JG, Green H (1975) Serial cultivation of strains of human epidermal keratinocytes: the formation of keratinizing colonies from single cells. Cell 6:331–343
15. Lim X, Tan SH, Koh WL, Chau RM, Yan KS, Kuo CJ, van Amerongen R, Klein AM, Nusse R (2013) Interfollicular epidermal stem cells self-renew via autocrine Wnt signaling. Science 342:1226–1230

Chapter 5

Isolation and Culture of Hair Follicle Dermal Sheath Mesenchymal Stromal Cells

Dongrui Ma, Seng-Teik Lee, and Alvin Wen Choong Chua

Abstract

To date, little is published on the characterization and therapeutic potential of human mesenchymal stromal cells (MSCs) derived from hair follicle dermal sheath (DS). We present protocols for the isolation and culture of human DS-MSCs starting with the use of a dissecting microscope to separate out dermal sheaths from hair follicles for trypsin digestion. We also present the protocols for the adipogenic, osteogenic, and chondrogenic differentiation of these DS-MSCs as we seek to harness these cells for potential applications in stem cell therapy and tissue engineering.

Key words Hair follicle, Dermal sheath, Mesenchymal stromal cells, MSCs, Adipogenic differentiation, Osteogenic differentiation, Chondrogenic differentiation, Stem cells, Tissue engineering

1 Introduction

Hair follicles contain discrete populations of mesenchymal and epithelial cells [1]. In the hair follicle, dermal papilla and dermal sheath cells have been shown to have powerful hair inductive properties, but the stem cell properties of the hair follicle dermal sheath have largely been overlooked. We have recently isolated fibroblast-like cells from human hair follicle dermal sheath [2] similar to work reported earlier [3]. These cells which we termed as dermal sheath mesenchymal stromal cells (DS-MSCs) exhibited proliferative capacity and colony-forming ability in in vitro culture. They also displayed immunophenotypic properties similar to those of bone marrow mesenchymal stromal cells (BM-MSCs), although with some variation. DS-MSCs have the ability to differentiate into adipocytes, chondrocytes, and osteoblasts and demonstrate high telomerase activity compared to BM-MSCs. The identification and characterization of these human hair follicle DS-MSCs provide a novel alternative source to BM-MSCs with potential applications in stem cell therapy and tissue engineering.

Sophie Böttcher-Haberzeth and Thomas Biedermann (eds.), *Skin Tissue Engineering: Methods and Protocols*, Methods in Molecular Biology, vol. 1993, https://doi.org/10.1007/978-1-4939-9473-1_5,

2 Materials

2.1 Solutions

1. 2.5 mg/mL Dispase II: Dissolve 500 mg of Dispase II powder in 200 mL of Dulbecco's modified Eagle's medium (DMEM) with gentle stirring. Sterile filter at 0.22 μm, dispense 10 mL of the dissolved solution in 15 mL centrifuge tubes, and store at 4 °C for short-term storage (up to a week). For long-term storage, keep at −20 °C (*see* **Note 1**).
2. 0.1% Collagenase–0.25% Dispase II mix: Warm the 10 mL tube of 2.5 mg/mL Dispase II to 37 °C prepared in **step 1**. Dissolve 10 mg of collagenase powder in the warmed Dispase II. Sterile filter the mix and it is ready for use to isolate cells from the hair follicle dermal sheath.
3. 0.1% Trypsin solution: Add 0.4 g glucose, 0.4 g trypsin powder, and 1.2 mL 0.5% phenol red into 400 mL of 1× phosphate-buffered saline (PBS). Adjust final pH to 7.45 just before filtration with up to 1 mL of 1 N NaOH (sodium hydroxide). Sterile filter at 0.22 μm and dispense into 50 mL centrifuge tubes. Store at –20 °C and thaw to 37 °C before use.
4. 0.02% Ethylenediaminetetraacetic acid (EDTA) solution: To prepare 2 L, add 0.4 g NaEDTA powder into 2000 mL 1× PBS. Sterile filter at 0.22 μm into 500 mL or 1 L bottles and they can be stored at room temperature.

2.2 Culture Medium

Sterilize all culture media after mixing all components using 0.22 μm filtration.

1. DS-MSC growth medium: 1% L-Glutamine, 20% fetal calf serum (FCS), and 1× penicillin/streptomycin (P/S) mix in DMEM.
2. Adipogenic differentiation medium: 1% L-Glutamine, 10% FBS, 0.5 mM isobutylmethylxanthine, 1 μM dexamethasone, 10 μM insulin, 200 μM indomethacin, and 1× penicillin/streptomycin (P/S) mix in DMEM.
3. Osteogenic differentiation medium: 10% FBS, 0.1 μM dexamethasone, 50 μM ascorbate-2-phosphate, and 10 mM β-glycerol phosphate, and 1× penicillin/streptomycin (P/S) mix in DMEM.
4. Chondrogenic differentiation medium: 1% FBS, 6.25 μg/mL insulin, 10 ng/mL TGF-β1, 50 nM ascorbate-2-phosphate, and 1× penicillin/streptomycin (P/S) mix in DMEM.

2.3 Staining and Fixation

1. 2% Crystal violet: Dilute 4 g crystal violet powder to 200 mL 95% methanol (*see* **Note 2**).
2. 0.7% Oil red O staining: Dissolve 0.7 g Oil red O in 100 mL propylene glycol, slowly stirring, heating to 100 °C for 10 min. Filter through Whatman #2 filter paper.

3. von Kossa staining (5% silver nitrate solution): Dissolve 5 g silver nitrate in 100 mL distilled water, mix well, store in clean brown bottle, and protect from light.
4. Alcian blue: Dissolve 1 g Alcian blue in 97 mL distilled water, add 3 g acetic acid, and mix well.
5. 10% Buffered formalin.

2.4 Equipment

1. Horizontal laminar flow hood with an embedded dissecting microscope.
2. Class II biosafety cabinet (BSC) for cell culture.
3. Chemical fume hood.
4. Phase-contrast microscope with digital imaging capture system.
5. CO_2 incubator with controlling and monitoring system for CO_2, humidity, and temperature.
6. Cell culture centrifuge.
7. Glass hemocytometer (with improved Neubauer slide).
8. Cell culture disposables: 30 g Needles, 1 mL syringes, 100 mm cell culture dishes, multiwell plates, 15 and 50 mL centrifuge tubes, pipetman, pipettes, micropipette, and pipette tips.

3 Methods

3.1 Isolation of Hair Follicle Sheath (HFS) Cells from Human Hair Follicle

1. Collect excess human scalp tissues with embedded hair follicles wrapped in sterile gauze and soaked with DMEM as transport medium in a sterile container (*see* **Note 3**).
2. Remove human scalp tissues from transport container and wash them in 100 mm cell culture dish with 20 mL of 1× PBS twice to wash off blood and debris in Class II BSC for cell culture.
3. Gently aspirate away all the PBS using a pipette.
4. Add 10 mL of fresh 1× PBS and gently spread the reagent out to wet the entire surface of the dish.
5. Bring covered 100 mm dish to dissecting microscope in horizontal laminar flow hood.
6. Remove the cover of 100 mm dish. Gently microdissect out each strand of hair follicle from the scalp using a 30 g needle attached to a 1 mL syringe under a dissecting microscope, ensuring that the dermal sheath is intact as shown in Fig. 1a.
7. After microdissecting out at least ten strands, transfer all these strands of hair follicle with intact dermal sheath into another empty 100 mm cell culture dish.

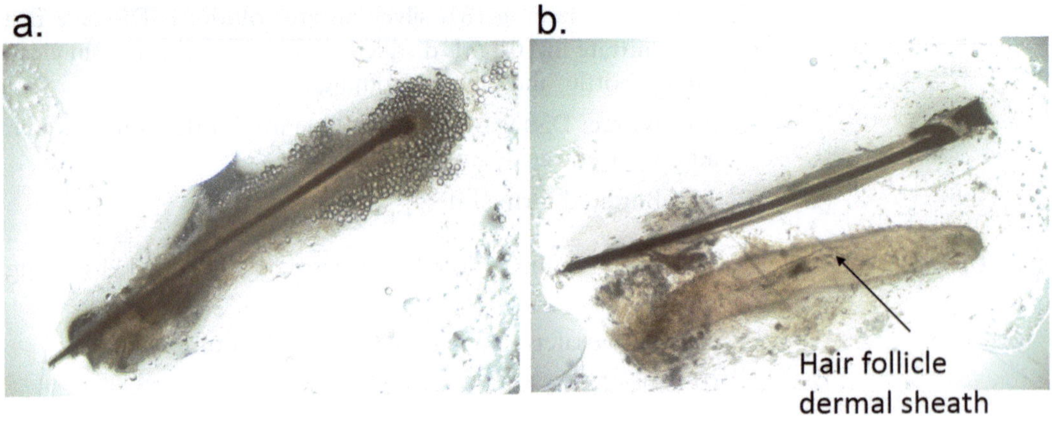

Fig. 1 (**a**) Single human hair follicle isolated from scalp tissue viewed under inverted microscope. Reproduced from Ma et al. [2] with permission from Elsevier. (**b**) The hair follicle dermal sheath that was carefully separated by microdissection away from the main shaft of the hair follicle after treatment with 0.1% collagenase–0.25% Dispase II. Reproduced from Ma et al. [2] with permission from Elsevier

8. Immediately transfer this covered 10 mm culture dish with hair follicles back into the BSC and aseptically add 10 mL of 0.1% collagenase–0.25% Dispase II mix into the dish.
9. Transfer this dish into the CO_2 incubator at 37 °C and incubate for 30 min.
10. Remove the dish from incubator and under the dissecting microscope in the laminar flow hood gently separate the dermal sheath from the hair follicle shaft with the 30 g needle, ensuring that the sheath is in one intact piece as shown in Fig. 1b.
11. Aspirate out all the 0.1% collagenase–0.25%Dispase II mix from the dish with a pipette.
12. Gently wash with 10 mL of 1× PBS three times and gently aspirate out the PBS (*see* **Note 4**).
13. Remove and discard all hair follicle shafts using sterile forceps.
14. Add 1 mL of 0.02% EDTA followed by 1 mL of 0.1% trypsin solution (all pre-warmed to 37 °C) into the dish.
15. Mince the sheath tissues with scalpel blades for 5 min and then incubate these minced tissues in the same 0.1%trypsin/0.02EDTA mix for further 30 min at 37 °C.
16. Add 1 mL of DS-MSC growth medium to quench the trypsinization process.
17. Collect cell suspension from the dish into a 50 mL centrifuge tube.
18. Repeat **steps 7–17** until all the hair follicles are dissected or until the centrifuge tube is full.

19. Centrifuge the cell suspension at 200 × *g* for 5 min.
20. Aspirate out the supernatant and resuspend with appropriate medium volume of DS-MSC growth medium (*see* **Note 5**).
21. Do cell count with hemocytometer to obtain cell density (number of cells per mL) and total number of cells based on the dilution volume above (*see* **Note 6**).
22. Seed 2×10^5 of the isolated cells in a 100 mm cell culture dish each (*see* **Note 7**).

3.2 Cell Culture of Isolated DS-MSCs

1. Change to fresh DS-MSC growth medium (10 mL for 100 mm plate) every 3 days and check the cultures daily (*see* **Note 8**).
2. When the DS-MSCs are about 90% confluent in 8–12 days, subculture the DS-MSCs using 5 mL of 0.02% EDTA and 5 mL 0.1% trypsin solution in the 100 mm cell culture dish.
3. Add 5 mL of DS-MSC growth medium to quench the trypsinization process.
4. Collect cell suspension from the dish into a 50 mL centrifuge tube.
5. Perform **steps 20** and **21** described in Subheading 3.1.
6. Do cell count with hemocytometer, and obtain cell density (number of cells per mL) and total number of cells.
7. Seed 3.5×10^5 of the isolated cells in a 100 mm cell culture dish each for further subculture or analysis (*see* **Note 9**).

3.3 Colony-Forming Efficiency (CFE) Assay of DS-MSCs

1. After either **step 21** in Subheading 3.1 or **step 6** in Subheading 3.2, do step dilution of cell suspension after cell counting.
2. Briefly, draw out 1 mL of cell suspension in fresh centrifuge tube and dilute to 10^5 cells per mL with DS-MSC growth medium.
3. Next draw out another 1 mL from the tube in the above step into another fresh centrifuge and add 9 mL of DS-MSC growth medium to obtain 10^4 cells per mL.
4. Repeat **step 3** above to obtain final concentration of 10^3 cells per mL.
5. Seed 200 μL of medium from the tube with 10^3 cells per mL into 100 mm culture dish (work in duplicate) with 10 mL of DS-MSC growth medium to start the CFE assay. This is considered day 0.
6. Change to fresh DS-MSC growth medium on days 4 and 8.
7. On day 12, aspirate all cell culture medium from 100 mm plate and add 6 mL of 10% buffered formalin to cover the culture surface of the dish for 30 min at room temperature to fix the cultures.

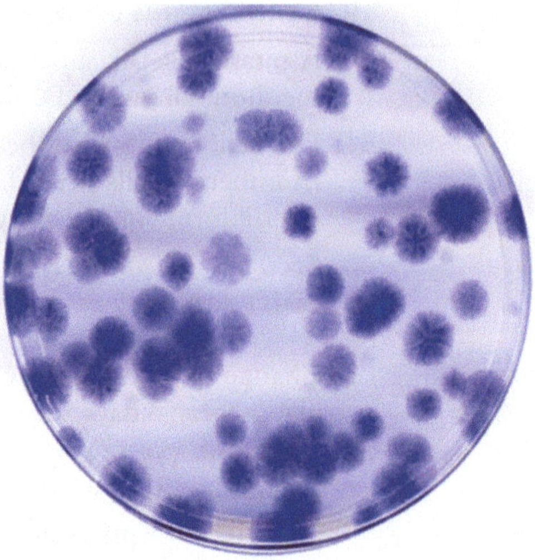

Fig. 2 A representative plate stained with 2% crystal violet solution after 12 days of culture with 200 DS-MSCs seeded

8. Aspirate out all the buffered formalin, and rinse the dish with distilled water.
9. Add 10 mL of 2% crystal violet solution for 15 min to stain the culture in the 100 mm culture dish.
10. Aspirate out the 2% crystal violet solution in hazardous waste bottle, rinse the cultures with distilled water, and allow the dish to dry. The staining will have the appearance as shown in Fig. 2.
11. The number of colonies will be counted under a dissecting microscope and the CFE is determined as percentage relative to the 200 cells seeded.

3.4 Induction of DS-MSCs Toward Adipogenic Differentiation

1. After **step 6** in Subheading 3.2, seed isolated DS-MSCs at a cell density of $5 \times 10^4/\text{cm}^2$ in 6-well culture plate and culture with 2 mL of DS-MSC growth medium in each well. Work in duplicates (*see* **Note 10**).
2. Allow the seeded DS-MSCs to proliferate and be confluent. This will occur typically between 1 and 3 days.
3. When cultures are confluent, change to 2 mL adipogenic differentiation medium each for the wells meant for induction and 2 mL of DS-MSC growth medium each for wells which serve as negative control.
4. Continue to change media every 3 days and culture the cells for 2 weeks.
5. After 2 weeks, aspirate away culture media and wash cultures two times with 1× PBS.

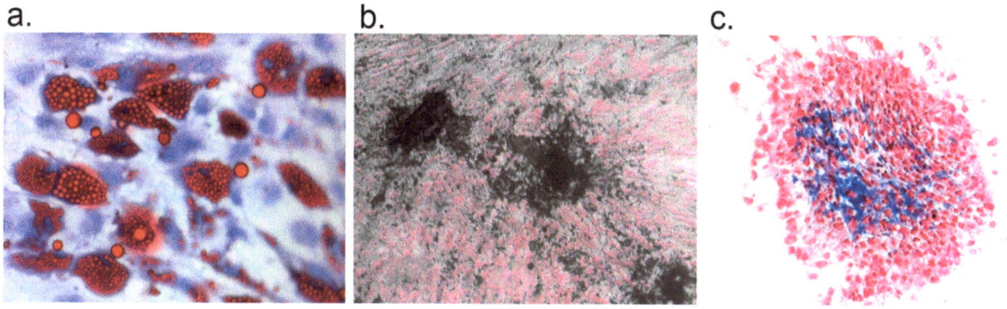

Fig. 3 In vitro differentiation of DS-MSCs. Reproduced from Ma et al. [2] with permission from Elsevier. (**a**) Positive Oil red O staining, indicating the presence of lipid vacuoles within the adipocytes after adipogenic induction of DS-MSCs. (**b**) Positive von Kossa staining which represented formation of black calcium after osteogenic induction of DS-MSCs. (**c**) Positive Alcian blue staining, indicating cartilage extracellular matrix formation after chondrogenic induction of DS-MSCs

6. Add 1.5 mL of 10% buffered formalin to cover the culture surface of each well with seeded DS-MSCs for 15 min at room temperature to fix the cultures.
7. Aspirate out all the buffered formalin and rinse the wells with distilled water.
8. Add 2 mL of 2% Oil red O staining solution to the wells for 30 min at room temperature (*see* **Note 11**).
9. Aspirate out the staining solution and wash the wells with 1× PBS twice.
10. Examine for generation of oil droplets in the cytoplasm of the differentiated cells under phase-contrast microscope (Fig. 3a).

3.5 Induction of DS-MSCs Toward Osteogenic Differentiation

1. After **step 6** in Subheading 3.2, seed isolated DS-MSCs at a cell density of $5 \times 10^4/\text{cm}^2$ in 6-well culture plate and culture with 2 mL of DS-MSC growth medium in each well. Work in duplicates.
2. Allow the seeded DS-MSCs to proliferate and be confluent. This will occur typically between 1 and 3 days.
3. When cultures are confluent, change to 2 mL osteogenic differentiation medium each for the wells meant for induction and 2 mL of DS-MSC growth medium each for wells which serve as negative control.
4. Continue to change media every 3 days and culture the cells for 2 weeks.
5. After 2 weeks, aspirate away culture media and wash cultures two times with PBS.
6. Add 1.5 mL of 10% buffered formalin to cover the culture surface of each well with seeded DS-MSCs for 15 min at room temperature to fix the cultures.

7. Aspirate out all the buffered formalin and rinse the wells with distilled water.
8. Add 2 mL of von Kossa staining solution for 30 min at room temperature (*see* **Note 11**).
9. Aspirate out the staining solution and wash the wells with 1× PBS twice.
10. Examine for mineral formation in the differentiated cells as indicted by the dark stains (Fig. 3b).

3.6 Induction of DS-MSCs Toward Chondrogenic Differentiation

1. After **step 6** in Subheading 3.2, seed isolated DS-MSCs at a cell density of $5 \times 10^4/\text{cm}^2$ in 6-well culture plate and culture with 2 mL of DS-MSC growth medium in each well. Work in duplicates.
2. Allow the seeded DS-MSCs to proliferate and be confluent. This will occur typically between 1 and 3 days.
3. When cultures are confluent, change to 2 mL chondrogenic differentiation medium each for the wells meant for induction and 2 mL of DS-MSC growth medium each for wells which serve as negative control.
4. Continue to change media every 3 days and culture the cells for 3–4 weeks.
5. After 2 weeks, aspirate away culture media and wash cultures two times with PBS.
6. Add 1.5 mL of 10% buffered formalin to cover the culture surface of each well with seeded DS-MSCs for 15 min at room temperature to fix the cultures.
7. Aspirate out all the buffered formalin and rinse the wells with distilled water.
8. Add 2 mL of Alcian staining solution for 30 min at room temperature (*see* **Note 11**).
9. Aspirate out the staining solution and wash the wells with 1× PBS twice.
10. Examine for cartilage extracellular matrix formation as indicted by the blue stains (Fig. 3c).

4 Notes

1. The Dispase II solution works best at neutral pH; therefore make sure that the Dispase II powder is dissolved in fresh DMEM to prevent elevated pH and make sure that the cap of the holding tubes is quickly closed after dispensing.
2. Methanol is hazardous and flammable. Dilute in chemical fume hood and wear appropriate personal protective equipment (PPE).

3. These tissues which would otherwise be discarded in plastic surgery hair transplantation procedures were collected with consent from patients and approval from the Institutional Review Board (IRB) of the hospital.
4. Beware not to aspirate out the hair follicles and the sheath pieces especially if vacuum pump is used.
5. Typically, 2–3 mL of growth medium is used to resuspend the DS-MSCs isolated from ten strands of hair follicle.
6. For more accurate cell count, keep average cell count number of the four corner squares (1 mm^2 each) to between 20 and 90 cells per square. Otherwise, readjust to appropriate dilution volume and repeat the cell count.
7. Cell yield direct from dermal sheath in primary culture is typically low and that is why seeding density is lower at 2×10^5 cells per 100 mm cell culture dish. Subsequent cell seeding density is 3.5×10^5 cells per 100 mm cell culture dish during cell passaging.
8. The DS-MSCs will start to form colonies with sparse density of cells within each colony (Fig. 4a), observable after 4–5 days and each colony will be densely packed in 10–12 days' time (Fig. 4b).
9. Freshly isolated, cultivated human DS-MSCs—characterized with flow cytometry analysis for cell surface phenotype—were found to be highly expressed for CD105, CD29, CD49b, and CD49d, while weakly expressed for CD90, CD44, CD34, and CD45 (Fig. 5).

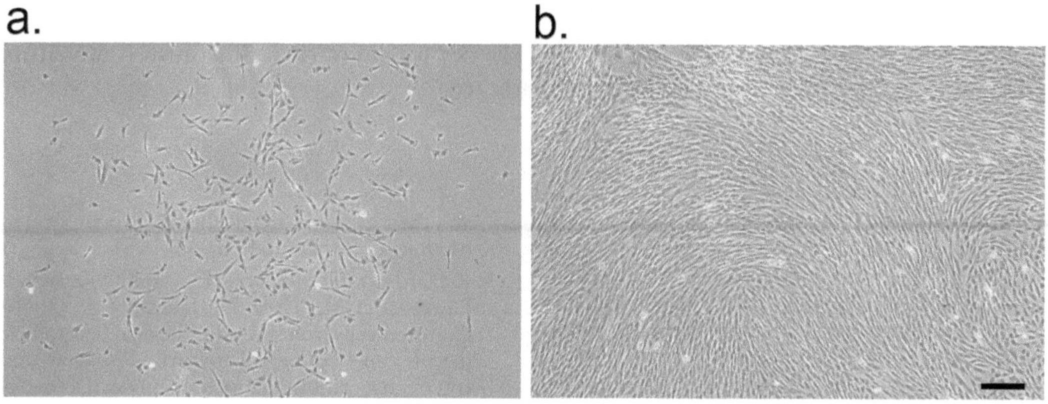

Fig. 4 Phase-contrast image of a colony HF dermal sheath mesenchymal stromal cells (DS-MSCs) after isolation and plated for (**a**) 4–5 days and (**b**) 10–12 days. Scale bar = 50 μm

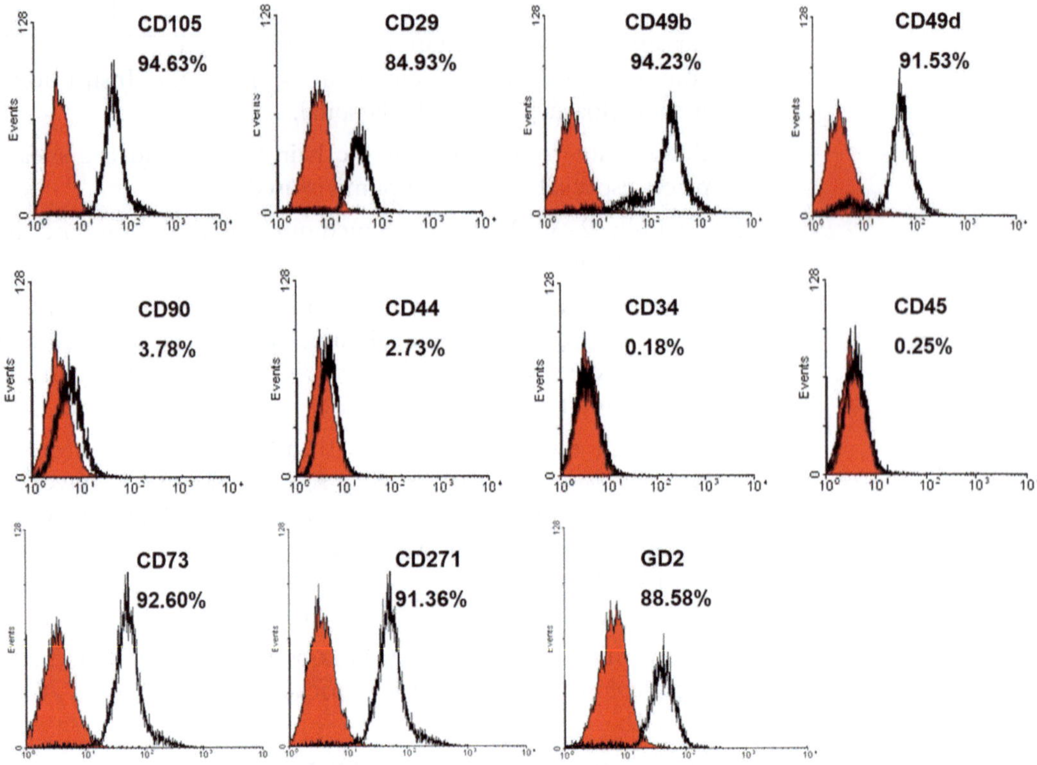

Fig. 5 Representative flow cytometry results of DS-MSCs stained with antibodies against CD105, CD29, CD49b, CD49d, CD90, CD44, CD34, and CD45. DS-MSCs (black peaks) strongly expressed in CD105, CD29, CD49b, and CD49d but not in CD90, CD44, CD34, and CD45. Isotype controls are shown in red peaks

10. In this case, working in duplicate means that two wells will be used to induce differentiation and two wells will serve as negative control.
11. Do not let the cells dry for more than 30 s during the entire staining procedure.

References

1. Richardson GD, Arnott EC, Whitehouse CJ, Lawrence CM, Reynolds AJ, Hole N et al (2005) Plasticity of rodent and human hair follicle dermal cells: implications for cell therapy and tissue engineering. J Investig Dermatol Symp Proc 10:180–183
2. Ma D, Kua JE, Lim WK, Lee ST, Chua AW (2015) In vitro characterization of human hair follicle dermal sheath mesenchymal stromal cells and their potential in enhancing diabetic wound healing. Cytotherapy 17:1036–1051
3. Richardson GD, Arnott EC, Whitehouse CJ, Lawrence CM, Hole N, Jahoda CA (2005) Cultured cells from the adult human hair follicle dermis can be directed toward adipogenic and osteogenic differentiation. J Invest Dermatol 124:1090–1091

Chapter 6

Isolation and Culture of Human Dermal Fibroblasts

Marta A. Kisiel and Agnes S. Klar

Abstract

Dermal fibroblasts are the main cell type present in skin connective tissue (dermis). Fibroblasts interact with epidermal cells during hair development and in interfollicular skin. Moreover, they play an essential role during cutaneous wound healing and in bioengineering of skin. Hence, culture of primary fibroblast is gaining in importance. In addition, fibroblasts established from skin biopsies provide a powerful tool for investigating normal skin physiology or specific disease states. In this chapter, detailed procedures for establishing and maintaining primary cultures of adult human dermal fibroblasts are described.

Key words Skin, Fibroblasts, Dermis, Collagen, Organotypic culture, Skin equivalent, Regenerative medicine, Tissue engineering

1 Introduction

Human skin fibroblasts are mesenchymal/stromal cells derived from the embryonic mesoderm. They are located in the dermal layer of skin, where they produce extracellular matrix proteins to strengthen the dermal compartment and interact with epidermal cells. The extracellular matrix synthesized by fibroblasts is rich in type I and/or type III collagen [1]. The mesenchymal–epithelial interactions regulate cell differentiation and pigmentation [2, 3], and maintain tissue homeostasis of the epidermis [4]. Dermal fibroblasts are characterized by their spindle morphology and expression of stromal specific markers such as fibronectin, vimentin, and TH-1 (CD90) [1]. However, fibroblasts from various organs, body sites, and spatial locations exhibit differences regarding marker expression and functional specialization [5–7].

Further, dermal fibroblasts are key players for orchestrating physiological tissue repair. During skin wound healing dermal fibroblasts switch from a proliferative and migratory phenotype to a contractile, matrix-remodeling myofibroblast [8, 9].

Myofibroblasts are identified on the basis of their ultrastructural morphology, with characteristic microfilament bundles in

Sophie Böttcher-Haberzeth and Thomas Biedermann (eds.), *Skin Tissue Engineering: Methods and Protocols*, Methods in Molecular Biology, vol. 1993, https://doi.org/10.1007/978-1-4939-9473-1_6,

their cytoplasm distinguishing them from quiescent dermal fibroblasts. Myofibroblasts also demonstrate fibronexus junctions with the surrounding extracellular matrix (ECM) showing some of the morphological characteristics of smooth muscle (SM) cells [10]. In addition, myofibroblasts regenerate connective tissue after an injury by collagen deposition and wound contraction. Afterwards, myofibroblasts disappear by apoptosis, thereby preventing excess contraction. However, under pathological conditions, excessive wound contraction may lead to fibrosis and scarring [11].

A promising development to improve wound healing and reduce scarring is the bioengineering of skin substitutes, which has made significant progress over the last few decades [12, 13].

The establishment of appropriate culture conditions for fibroblasts paved the way for a successful application of those cells in bioengineered skin substitutes. The addition of dermal fibroblasts improved consistently graft take and reduced susceptibility to infection and graft contracture as compared to the previously used epidermal sheets, such as autologous cultured epidermal autografts (CEA) [14].

Furthermore, the addition of the dermal component to the skin substitutes resulted in a more stable skin graft with better functional and esthetic properties [12].

For several years, our laboratory has developed and extensively characterized dermo-epidermal skin substitutes (DESS) to treat full-thickness skin defects in clinical applications [3, 15–19].

Most recently, we reconstituted pigmented dermo-epidermal skin analogs from different pigmentation types and achieved very satisfactory in vivo results restoring the patients' native skin color [19].

To summarize, our 3-dimensional DESS offer an exceptional model to study cell-cell interactions, as we demonstrated, e.g., the importance of different fibroblast populations in the context of reproducing the donor skin color in our pigmented skin substitutes [2, 3], as well as specific mechanisms involved in pathologies such as fibrosis [8, 9].

This chapter describes methods and materials needed for all aspects of fibroblast culture, namely tissue preparation, isolation, cell culture, passaging, and cryopreservation.

2 Materials

2.1 Media and Additives

1. DMEM medium: Dulbecco's modified Eagle's medium (DMEM), containing 110 mg/L sodium pyruvate and 862 mg/L L-glutamine; store in the dark at 4 °C.
2. Fetal calf serum.

Thaw sera at 4 °C or in cold water (*see* **Note 1**). Gently swirl it to resuspend its components. Inactivate in a 56 °C water bath for 30 min and distribute in single-use aliquots. Store at −20 °C or at −80 °C for long-term storage.

3. Penicillin/streptomycin: 50000 IU/mL Penicillin/streptomycin (according to the manufacturer's recommendations).
4. Hepes (4-(2-hydroxyethyl)-1-piperazineethanesulfonic acid) buffer: 1 M Stock solution.
5. Fungizone: 0.25 mg/mL Amphotericin B in ultrapure water to make a 500× stock solution.
6. Gentamycin: 10 mg/mL.

2.2 Complete Media

1. Tissue transport medium (tDMEM): 500 mL DMEM, 15 mL 0.75 μg/mL fungizone, 15 mL 300 U/mL penicillin, 15 mL 300 μg/mL streptomycin, 0.75 mL 15 μg/mL gentamycin; store in the dark at 4 °C for up to 6 weeks.
2. Complete fibroblast culture medium (DMEM complete): 500 mL DMEM, 10% v/v of fetal calf serum, 5 mL 10,000 U/mL penicillin/10 mg/mL streptomycin, 5 mL 1 M HEPES; store in the dark at 4 °C for up to 4 weeks.
3. Cryopreservation medium: Add 10% v/v of dimethyl sulfoxide (DMSO) to fetal calf serum, keep on ice or store at 4 °C, and use within the day.

2.3 Other Solutions

1. Phosphate-buffered saline (PBS) without Ca^{2+}/Mg^{2+}: Store at room temperature.
2. Collagenase solution: 10 mg Collagenase, 1 mL DMEM/10% FCS, 4 mL DMEM w/o, sterilize by filtration through a 0.22 μm low-binding disposable filter.
3. Dispase solution: 2.5 mL 5000 U/mL Dispase, 2.5 mL PBS, 250 μL 10 mg/mL gentamycin.
4. Trypsin/EDTA: 0.05% w/v Trypsin, 0.01% w/v EDTA, store at −20 °C to −80 °C.
5. Trypan blue solution: 0.4% Trypan blue in distilled water.

2.4 Tissues and Cells

1. Normal human fibroblasts (NHF): Isolation from surgically removed 3 to 6 cm^2 skin biopsies.

2.5 Materials

1. 0.22 μm Low-binding disposable filters.
2. 100 μm Strainers.
3. Sterile containers.
4. 50 mL Tubes.
5. 100 × 15 mm Cell culture Petri dishes.

6. Dissecting curved forceps.
7. Size 3 scalpel and size 22 blade.
8. 25 or 75 cm^2 tissue culture flasks.
9. Sterile cryogenic vials.
10. Freezing containers.

3 Methods

Manipulations must be performed under a sterile laminar flow cabinet.

3.1 Tissue Sampling and Transport

1. Immediately following aseptic (*see* **Note 2**) surgical removal, put the skin biopsy into a sterile container filled with cold (4 °C) tDMEM.
2. Samples must be kept at 4 °C (*see* **Note 4**).

3.2 Isolation

1. With sterile forceps, transfer the skin sample into a 100 × 15 mm cell culture Petri dish.
2. Use curved forceps to gently spread out the skin sample with the epidermis facing top in a Petri dish.
3. Use a scalpel with a size 22 blade to cut the sample into small pieces of approximately 4–6 mm diameter and transfer into a 50 mL tube.
4. Add 2.5 mL of Dispase solution.
5. Incubate overnight at 4 °C.
6. Use two curved forceps to carefully separate the dermis from the epidermis.
7. Incubate with Collagenase solution for 60 min at 37 °C, 5% CO_2; shake regularly.
8. Add 15 mL of DMEM complete medium into a 50 mL Falcon (*see* **Note 3**).
9. Pass through a 100 μm strainer.
10. Centrifuge (200 × *g*) the NHF suspension for 5 min at room temperature.
11. Resuspend cells in 10 mL of DMEM complete medium, stain with trypan blue solution, and count NHF using an automated cell counter or a hemocytometer (*see* **Note 5**).

    Cell viability is expected to be greater than 80%.
12. Seed approximately 1–3 × 10^6 NHF into an uncoated 75 cm^2 tissue flask in DMEM complete medium. Total medium volume should not exceed 15 mL per 75 cm^2 (*see* **Note 6**).

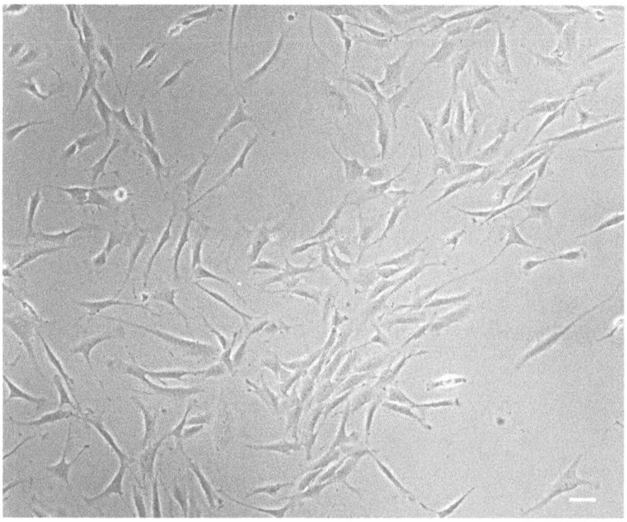

Fig. 1 First passage of NHF at 10× under a phase-contrast microscope. The NHF confluence is estimated at 70%. Bar is 50 μm

3.3 Culture

1. Incubate NHF in an 8% CO_2 and 100% humidity atmosphere at 37 °C (*see* **Note** 7, Fig. 1).
2. Change the culture medium three times a week, every 2–3 days. Remove the medium from the culture flask. Replace it with warm (37 °C) DMEM complete medium.
3. Monitor NHF morphology and confluence under a microscope (*see* **Note 8**).
4. When NHF reach 70–90% confluence, subculture (*see* Subheading 3.4) or cryopreserve (*see* Subheading 3.5) them.

3.4 Subculture (Passage)

1. Remove the culture medium.
2. Rinse NHF in 75 cm^2 culture flasks with 5 mL PBS w/o Ca^{2+}/ Mg^{2+}. Remove it.
3. Add 5 mL of trypsin/EDTA diluted 1:5 in PBS (4 mL trypsin + 1 mL PBS) into the 75 cm^2 culture flask.
4. Incubate for 5 min at 37 °C until all NHF are completely detached from the flask (verify cell detachment under a microscope) (*see* **Note 9**).
5. Neutralize trypsin activity by adding 5 mL of DMEM complete medium, and rinse rest from plate.
6. Vigorously pipette the NHF suspension up and down at least five times to ensure suspension homogeneity.
7. Transfer the NHF suspension into a 15 mL tube.
8. Centrifuge (200 × *g*) the NHF suspension for 5 min at room temperature (*see* **Note 10**).

9. Remove the supernatant and resuspend NHF in DMEM complete medium.
10. Count cells using an automated cell counter or a hemocytometer. Use trypan blue staining to estimate cell viability. Cell viability is expected to be greater than 95%.
11. Seed NHF at 1–3 × 10^6 cells per 75 cm^2 or more into an uncoated culture flask. Seed directly into the culture medium. Use 10 mL of total medium volume per 75 cm^2.

3.5 Cryopreservation

1. Fill a freezing container with 100% isopropyl alcohol and store it at 4 °C until cooled.
2. Follow **steps 1** through **11** from Subheading 3.4.
3. Remove the supernatant and resuspend NHF at the desired concentration in cryopreservation medium. Put the tube on ice.
4. Aliquot in cryogenic vials on ice.
5. Put the cryogenic vials in the freezing container.
6. Store the container overnight at −80 °C. In these conditions, cell temperature should drop 1 °C/min.
7. Store cryogenic vials in liquid nitrogen for long-term storage.

3.6 Thawing

1. Put the cryogenic vial in a 37 °C water bath. Do not let the NHF suspension thaw completely. A small ice pellet should remain (*see* **Note 11**).
2. Transfer the cell suspension immediately from the cryogenic vial into a 50 mL tube containing 5–10 mL of warm (37 °C) DMEM complete medium.
3. Rinse the cryogenic vial with 1–1.5 mL of the suspension.
4. Centrifuge (200 × *g*) the NHF suspension for 5 min at room temperature.
5. Remove the supernatant and resuspend NHF in 10 mL of warm (37 °C) DMEM complete medium.
6. Use an automated cell counter or a hemocytometer to count the NHF.
7. Use trypan blue staining to estimate cell viability. Cell viability is expected to be greater than 80%.
8. Centrifuge (200 × *g*) the NHF suspension for 5 min at room temperature.
9. Remove the supernatant and resuspend NHF at the desired concentration in DMEM complete medium.
10. Seed NHF at 1–3 × 10^6 cells per 75 cm^2 or more into an uncoated culture flask. Seed directly into the culture medium. Use 10 mL of total medium volume per 75 cm^2.

4 Notes

1. Sera and other frozen components can also be thawed more rapidly at room temperature or in a 37 °C water bath. However, this procedure is not recommended.
2. Adult skin is contaminated with (mainly) nonpathogenic bacteria or molds. Therefore, these microorganisms must be removed prior to skin biopsy to insure good sterility of the tissue.
3. Culture media are supplemented with antibiotics and antimycotics. However, they should be rinsed with 70% ethanol after prewarming to 37 °C, in water bath, to prevent contamination.
4. To achieve maximum cell yield and viability, skin tissue should be processed directly after surgery. However, when necessary, tissue can be kept at 4 °C for up to 5–7 days without significant loss of viability.
5. The yield of cells per cm^2 skin is affected by the age and health of donor. Younger donors generate more stem cells than older ones.
6. Ensure proper distribution of cells by quickly moving the flask back and forth several times. Open and close the incubator door gently after cell plating.
7. There may be only a few adherent cells after 24 h. Wait for at least 2 days to allow cells to attach to the bottom of the flask.
8. It is helpful to examine routinely cultures under microscope to determine their status and health. It is also recommended to record the details concerning cell cultures, including the subculture ratios and the passage number.
9. For optimal trypsinization efficiency, do not stack culture flasks on top of each other in the incubator. Temperature is typically higher on the flask surface, that are closest to the metal shelves.
10. Centrifugation following trypsin detachment of cells removes trypsin and cell debris. This removal helps to maintain viability and longevity of cultures.
11. DMSO is a toxic oxidative agent at temperatures above 10 °C. Therefore, work with this agent quickly.

References

1. Alberts B, Johnson A, Lewis J et al (2002) Fibroblasts and their transformations: the connective-tissue cell family. In: Molecular biology of the cell, 4th edn. Garland Science, New York
2. Biedermann T, Bottcher-Haberzeth S, Klar AS, Widmer DS, Pontiggia L, Weber AD, Weber DM, Schiestl C, Meuli M, Reichmann E (2015) The influence of stromal cells on the pigmentation of tissue-engineered dermo-

epidermal skin grafts. Tissue Eng Part A 21 (5–6):960–969

3. Klar AS, Biedermann T, Michalak K, Michalczyk T, Meuli-Simmen C, Scherberich A, Meuli M, Reichmann E (2017) Human adipose mesenchymal cells inhibit melanocyte differentiation and the pigmentation of human skin via increased expression of TGF-beta1. J Invest Dermatol 137 (12):2560–2569
4. Yamaguchi Y, Hearing VJ, Itami S, Yoshikawa K, Katayama I (2005) Mesenchymal-epithelial interactions in the skin: aiming for site-specific tissue regeneration. J Dermatol Sci 40(1):1–9
5. Driskell RR, Watt FM (2015) Understanding fibroblast heterogeneity in the skin. Trends Cell Biol 25(2):92–99
6. Lynch MD, Watt FM (2018) Fibroblast heterogeneity: implications for human disease. J Clin Invest 128(1):26–35
7. Philippeos C, Telerman SB, Oules B, Pisco AO, Shaw TJ, Elgueta R, Lombardi G, Driskell RR, Soldin M, Lynch MD, Watt FM (2018) Spatial and single-cell transcriptional profiling identifies functionally distinct human dermal fibroblast subpopulations. J Invest Dermatol 138 (4):811–825
8. Tomasek JJ, Gabbiani G, Hinz B, Chaponnier C, Brown RA (2002) Myofibroblasts and mechano-regulation of connective tissue remodelling. Nat Rev Mol Cell Biol 3 (5):349–363
9. Darby IA, Hewitson TD (2007) Fibroblast differentiation in wound healing and fibrosis. Int Rev Cytol 257:143–179
10. Eyden B (2005) The myofibroblast: a study of normal, reactive and neoplastic tissues, with an emphasis on ultrastructure. Part 2—tumours and tumour-like lesions. J Submicrosc Cytol Pathol 37(3–4):231–296
11. Gabbiani G (2003) The myofibroblast in wound healing and fibrocontractive diseases. J Pathol 200(4):500–503
12. Biedermann T, Boettcher-Haberzeth S, Reichmann E (2013) Tissue engineering of skin for wound coverage. Eur J Pediatr Surg 23 (5):375–382
13. Klar AS, Zimoch J, Biedermann T (2017) Skin tissue engineering: application of adipose-derived stem cells. Biomed Res Int 2017:9747010
14. Wood FM, Kolybaba ML, Allen P (2006) The use of cultured epithelial autograft in the treatment of major burn injuries: a critical review of the literature. Burns 32(4):395–401
15. Braziulis E, Diezi M, Biedermann T, Pontiggia L, Schmucki M, Hartmann-Fritsch F, Luginbuhl J, Schiestl C, Meuli M, Reichmann E (2012) Modified plastic compression of collagen hydrogels provides an ideal matrix for clinically applicable skin substitutes. Tissue Eng Part C Methods 18(6):464–474
16. Pontiggia L, Biedermann T, Meuli M, Widmer D, Bottcher-Haberzeth S, Schiestl C, Schneider J, Braziulis E, Montano I, Meuli-Simmen C, Reichmann E (2009) Markers to evaluate the quality and self-renewing potential of engineered human skin substitutes in vitro and after transplantation. J Invest Dermatol 129(2):480–490
17. Klar AS, Guven S, Biedermann T, Luginbuhl J, Bottcher-Haberzeth S, Meuli-Simmen C, Meuli M, Martin I, Scherberich A, Reichmann E (2014) Tissue-engineered dermo-epidermal skin grafts prevascularized with adipose-derived cells. Biomaterials 35(19):5065–5078
18. Boettcher-Haberzeth S, Klar AS, Biedermann T, Schiestl C, Meuli-Simmen C, Reichmann E, Meuli M (2013) "Trooping the color": restoring the original donor skin color by addition of melanocytes to bioengineered skin analogs. Pediatr Surg Int 29(3):239–247
19. Boettcher-Haberzeth S, Biedermann T, Pontiggia L, Braziulis E, Schiestl C, Hendriks B, Eichhoff OM, Widmer DS, Meuli-Simmen C, Meuli M, Reichmann E (2013) Human eccrine sweat gland cells turn into melanin-uptaking keratinocytes in dermo-epidermal skin substitutes. J Invest Dermatol 133(2):316–324

Chapter 7

Isolation and Culture of Human Dermal Microvascular Endothelial Cells

Jennifer Bourland, Dominique Mayrand, Nathalie Tremblay, Véronique J. Moulin, Julie Fradette, and François A. Auger

Abstract

Primary endothelial cells are needed for angiogenesis studies, and more particularly in the field of tissue engineering, to engineer pre-vascularized tissues. Investigations often use human umbilical vein endothelial cells due to their extensive characterization, but also because they are easy to obtain and isolate. An alternative is the use of human dermal microvascular endothelial cells, more representative of adult skin angiogenesis and vascularization processes. This chapter presents a detailed methodology to isolate and culture microvascular endothelial cells from skin biopsies based on enzymatic digestion and mechanical extraction.

Key words Skin engineering, Endothelial cells, Microvascularization, Capillary, Cell isolation, Angiogenesis, Immunoselection

1 Introduction

Endothelial cells are widely used to investigate angiogenesis, but also to pre-vascularize biomaterials or tissue-engineered constructs. Different human endothelial cells are available for these studies, originating either from large-diameter vessels or capillaries. Many experiments use human umbilical vein endothelial cells (HUVEC), which are easy to isolate and culture, but these cells differ from native skin endothelial cells, in particular by the size of the vessels they can form in vitro and by their interactions with extracellular matrix components [1]. For example, human microvascular endothelial cells (HMVEC) are in direct contact with a basement membrane composed of proteins such as collagen type IV and laminin and thus are more prone to the production of these molecules compared to HUVEC. They are also more responsive to factors modulating angiogenesis such as angiopoietins [2]. HMVEC can be extracted from either adult or newborn skin using a method

Sophie Böttcher-Haberzeth and Thomas Biedermann (eds.), *Skin Tissue Engineering: Methods and Protocols*, Methods in Molecular Biology, vol. 1993, https://doi.org/10.1007/978-1-4939-9473-1_7,

based on enzymatic digestion and extrusion. The epidermis is first removed after enzymatic digestion. The capillaries are then isolated by mechanical extrusion from the dermis and seeded as explants. Colonies of endothelial cells can develop from endothelial cells originating from the tubules. The resulting population needs to be subsequently purified to remove residual fibroblasts, usually based on selection methods using CD31 expression. Indeed, CD31 (or platelet endothelial cell adhesion molecule-1, PECAM-1) is a molecule expressed at the surface of both blood and lymphatic microvascular endothelial cells [3], allowing their adhesion. It also mediates angiogenic processes through the formation of filopodia during vessel sprouting [4, 5]. Two main methods of purification are available using immunomagnetic beads [6] or flow-assisted cell sorting (FACS) [3]. This chapter details an immunomagnetic positive method to select endothelial cells.

2 Materials

Use ultrapure deionized and apyrogenic water to prepare the solutions and to rinse the tools before sterilization. Store all reagents at 4 °C unless indicated otherwise. Properly dispose of all materials which came in contact with skin samples and primary cells as they are considered a biolevel two safety hazard. All work needs to be performed under aseptic conditions, using sterile solutions and materials.

2.1 Cell Extraction and Culture

1. Skin resection: Human skin samples must be obtained with the consent of the donor or of his/her legal representative. The research project must have been previously approved by your institutional ethic committee.
2. Wash buffer: Dissolve 80 g NaCl, 2 g KH_2PO_4, 14.4 g Na_2HPO_4, and 2 g KCl in 1 L water for 10× stock solution; adjust the pH between 7.4 and 7.6; dilute with water to 1× working solution; ensure complete dissolution; filter solution with a 0.22 μm membrane under a biological hood; sterile solution can be stored at 4 °C for up to a month; right before use, add 100 U/mL penicillin G, 25 μg/mL gentamicin sulfate, and 0.5 μg/mL fungizone.
3. HEPES (4-(2-hydroxyethyl)-1-piperazineethanesulfonic acid) solution: 0.01 M HEPES, 1 mM $CaCl_2$, adjust the pH to 7.45, filter with a 0.22 μm membrane.
4. Hanks' balanced salt solution (HBSS): Dissolve 0.98 g of HBSS powder in 100 mL H_2O, supplement with 35 mg of sodium bicarbonate, adjust the pH to 7.4, and filter with a 0.22 μm membrane.

5. Dispase: Prepare 1% weight/volume dispase solution in HBSS, add HyClone™ FetalClone™ II Serum (H) to a final concentration of 2%, and prepare 10 mL for 12 cm^2 of skin or less.
6. EGM™-2MV: Use commercially available endothelial medium such as EGM™-2MV. Prepare it freshly by adding the SingleQuot kit supplements to the EBM-2 basal medium. The 5% FBS concentration adequately supports endothelial cell growth. The antibiotic in the SingelQuot Kit should be replaced by 100 U/mL penicillin G and 25 μg/mL gentamicin sulfate.
7. Sterile Petri dishes: At least five 100 mm Petri dishes and one 60 mm Petri dish for each 12 cm^2 of skin. They should not be tissue culture treated to prevent cell adhesion.
8. Two sterile scalpels (size number 4) with blades (size #22).
9. Two sterile stainless steel curved tweezers.
10. A set of sterile stainless steel long tweezers.
11. Sterile curved scissors.
12. Parafilm®.
13. 50 mL Conical sterile polypropylene tubes.
14. Gelatin-coated flasks: 0.1% Gelatin solution and filter with a 0.22 μm membrane. Add 4 mL of the gelatin solution to a 25 cm^2 T-flask or 10 mL to a 75 cm^2 T-flask and be careful to spread the gelatin over the entire flask area. Incubate at 4 °C overnight. Be careful to place the flasks on a flat surface to ensure homogeneity of the coating. The next day, aspirate the gelatin solution, without touching the coated surface. Coated flasks can be stored for 1 month at 4 °C.
15. Dulbecco's modified Eagle medium (DMEM) supplemented with 10% FBS and antibiotics (100 U/mL penicillin G and 25 μg/mL gentamicin sulfate).

2.2 Endothelial Cell Selection

1. Dynabeads™ CD31 Endothelial Cell: The number of beads needed can vary, but processing 12 cm^2 of skin usually requires 1–5 million beads.
2. DynaMag™-5 Magnet.
3. 5 mL Round-bottom polystyrene tubes compatible with the DynaMag™-5 Magnet.
4. HBSS with 2% FBS (*see* Subheading 2.1, **item 4**).
5. PBS supplemented with 0.1% (weight/volume) bovine serum albumin (BSA).
6. Ice.
7. Orbital shaker.
8. 15 mL Conical sterile polypropylene tubes.

2.3 Cell Amplification and Banking

1. Gelatin-coated culture dishes (*see* Subheading 2.1, **item 14**).
2. EGM™-2MV medium (*see* Subheading 2.1, **item 6**).
3. Cryopreservation medium (FBS with 10% dimethyl sulfoxide (DMSO)): To prepare 10 mL of cryopreservation medium, mix 9 mL of cold FBS with 1 mL of DMSO. Be careful to use cold solutions while freezing cells.
4. Cryogenic storage tubes.
5. Freezing container.

3 Methods

Carry out all steps under a biological hood and under sterile conditions unless otherwise specified. Skin samples should be collected and transported in DMEM supplemented with 10% FBS, 100 U/mL of penicillin G, and 25 μg/mL gentamicin sulfate. Keep at 4 °C until processing (maximum 48 h).

3.1 Day 1: Skin Sample Preparation

1. Remove the sample from the storage buffer in a sterile 100 mm Petri dish. A smaller Petri dish can be used for small samples.
2. Inspect the general aspect on both sides and note any particular characteristic (*see* **Note 1**).
3. Then, with the dermis facing upwards, remove any remaining adipose tissue with curved scissors. If damaged zones are present, they should be removed as well using a scalpel.
4. With clean sterile tweezers, retrieve the sample and place it in a new sterile Petri dish. The use of new sterile dishes or instruments contributes to reduce the risks of contamination. Place the epidermis on top and trace the shape of the skin to calculate its surface area (*see* Fig. 1).
5. Using long tweezers, transfer the sample to a 50 mL tube containing 40 mL of wash buffer. Depending of the sample size, add enough PBS to bring the total volume in the tube to 45 mL.
6. Add Parafilm® around the cap to prevent leakage.
7. Mix vigorously by shaking the tube for 30 s.
8. Discard the PBS by inversion and add a fresh PBS solution.
9. Repeat **steps 6–8** with a new wash buffer nine times (*see* **Note 2**).
10. Transfer pieces of skin into a sterile petri culture dish using new sterile forceps and cut the skin in pieces of approximatively 0.3 cm × 1 cm using a scalpel.

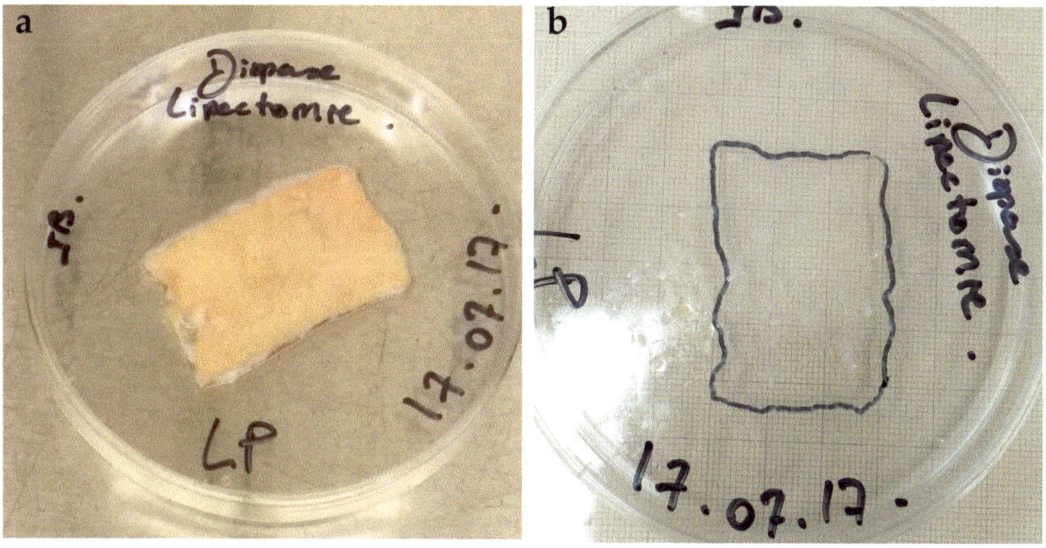

Fig. 1 Appearance of skin sample and surface calculation. (**a**) Skin sample appearance once spread in a Petri dish and (**b**) example of how to trace the corresponding border to measure skin surface. Knowing the tissue area allows to calculate cell yield per cm^2

11. Recommended: Take one piece of the skin to embed in optimal cutting temperature (OCT) compound for further characterization (*see* **Note 3**).
12. The next steps are optimized for 12 cm^2 of skin or less. If a larger sample is processed, duplicate the process in separate dishes and adapt all the following steps.
13. Transfer the pieces of skin in a new 60 mm Petri dish.
14. Add 10 mL of 1% dispase solution for 12 cm^2 of skin or less. Be careful to cover the whole sample (*see* **Note 4**).
15. Carefully seal the Petri dish with Parafilm® and place it inside a 100 mm Petri dish to avoid spillage and surface contamination with biological material.
16. Incubate overnight in the refrigerator (4–8 °C).

3.2 Day 2: Capillary Extrusion

This protocol is also designed for a maximum skin surface of 12 cm^2. If the skin is larger, divide samples in order to process 12 cm^2 of skin or less. Repeat the procedure with the rest of the skin samples.

1. After the overnight incubation, transfer one piece of skin into a new 60 mm Petri dish.
2. While maintaining the dermis in place with tweezers, use thin curved tweezers to delicately take hold of the epidermis and slowly peel it off the dermis. Place the epidermis in a sterile

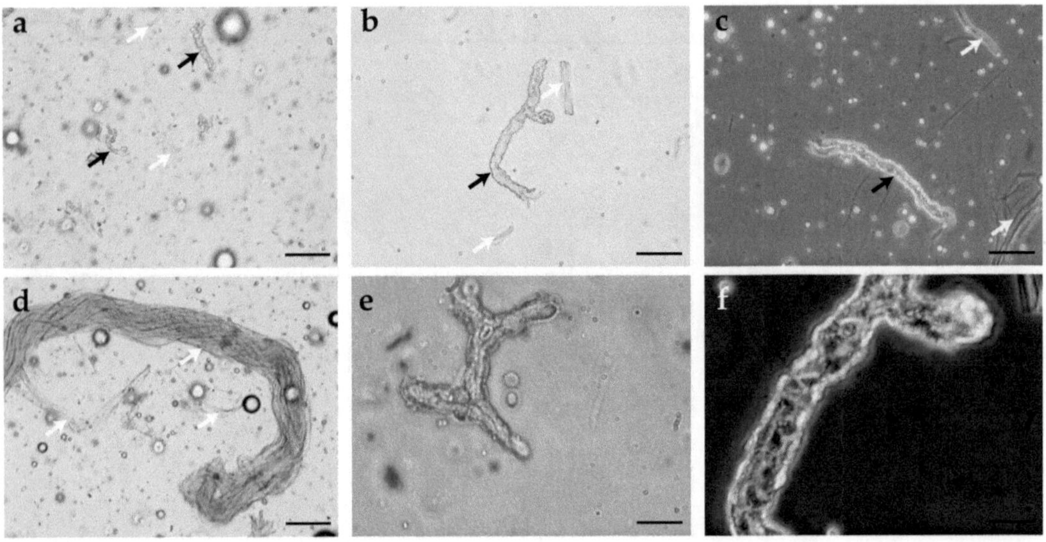

Fig. 2 Aspect of tubules obtained after extrusion. Light-phase microscopy aspect of tubules during the extrusion phase. (**a–f**) Many cells and tissue fragments are visible. (**a–d**) Tubules are indicated by black arrows, whereas extracellular matrix fragments are indicated by white arrows. (**b**, **c**, **e**, **f**) Different views of extracted tubules featuring a linear or branching morphology. Scale bars: (**a**, **d**) 250 μm, (**b**, **c**) 100 μm, (**e**) 25 μm, (**f**) 10 μm

Petri dish containing some DMEM medium if you want to extract the keratinocytes (*see* **Note 5**); otherwise discard it.

3. Place the dermis in a 100 mm Petri dish containing 25 mL of DMEM with 10% FBS and antibiotics. Repeat with each skin piece.
4. Transfer one piece of dermis in a 60 mm Petri dish containing 2 mL of EGM™-2MV and extrude the capillaries by applying regular pressure on the dermis using the curved part of the instrument and sliding toward the outside of the piece. The efficacy of the extrusion process can be verified using an optic microscope, as illustrated in Fig. 2.
5. After having confirmed the efficacy of your technique for extruding capillaries, proceed similarly for each piece of the dermis. (If you need to isolate the fibroblasts, keep the dermis for further processing after extrusion; *see* **Note 6**).
6. With a pipette, transfer the capillary and cell suspension obtained by extrusion in a 15 mL conical tube.
7. Centrifuge for 10 min at 300 × *g*.
8. Resuspend the pellet, and if possible count them with a hemocytometer (*see* **Note 7**).
9. Seed 1×10^6 cells per 25 cm^2 flask coated with gelatin (usually seed in 25 cm^2 for a foreskin, and in 10–25 cm^2 for 12 cm^2 of adult skin).

10. Incubate at 37 °C and 5–8% CO_2.
11. The following day, rinse the previously seeded cells up to three times with 5 mL EGM™-2MV (for 25 cm^2). This allows to detach dermal pieces and fibroblasts which are weakly attached. Maintain the cells in culture with 5 mL EGM™-2MV (for 25 cm^2). Change medium three times a week (every 48 h or 72 h).
12. In the subconfluent cultures, fibroblast contamination can be prevented by elimination of colonies of fibroblasts: if possible, identify fibroblast colonies located far enough from neighboring endothelial colonies using a microscope. Next, using Pasteur pipettes with an aspiration system, remove the supernatant and then aspirate carefully the fibroblast colony. Add 5 mL EGM-2MV and check for the presence of remaining fibroblasts with a microscope. Repeat if necessary (*see* **Note 8**).

3.3 Endothelial Cell Selection/Isolation

Endothelial cell purification can be performed either using immunomagnetic beads or by FACS using a directly coupled antibody directed against surface markers such as CD31 (positive selection). Immunomagnetic purification will be detailed in **steps 3–17**.

1. Remove culture medium and rinse with 1 mL of trypsin per 25 cm^2 flask. Add 3 mL of trypsin to the cells and incubate at 37 °C until complete cell detachment (3 mL for 25 cm^2 and 8 mL for 75 cm^2 flask).
2. After detachment, inhibit trypsin activity with the same volume of EGM-2MV. Transfer cells into a 15 mL conical sterile tube. Centrifuge cells at 300 × *g* for 10 min and perform a cell count.
3. To calculate the bead concentration needed, consider the proportion of fibroblasts in the culture (Fig. 3). If endothelial cells are prominent and appear as many large colonies (*see* Fig. 3f), consider a ratio of one bead for one cell. If more fibroblasts are present (*see* Fig. 3e), consider reducing the number of beads to a ratio between 1:2 and 1:4 beads per total cells (*see* **Note 9**).
4. Pipette a minimum of 25 μL of beads (equivalent to 1×10^7 beads) and place them in a 5 mL round-bottom polystyrene tube containing 2 mL of PBS 0.1% BSA.
5. Mix carefully and place the tube in the magnetic support to trap the beads. Incubate for 2 min.
6. While in the support, discard the supernatant by aspiration or pipetting. Be careful not to disturb the beads accumulated along the magnet.
7. Rinse twice with PBS-0.1% BSA.
8. Caution: The following steps need to be performed on ice with cold solutions (*see* **Note 10**).

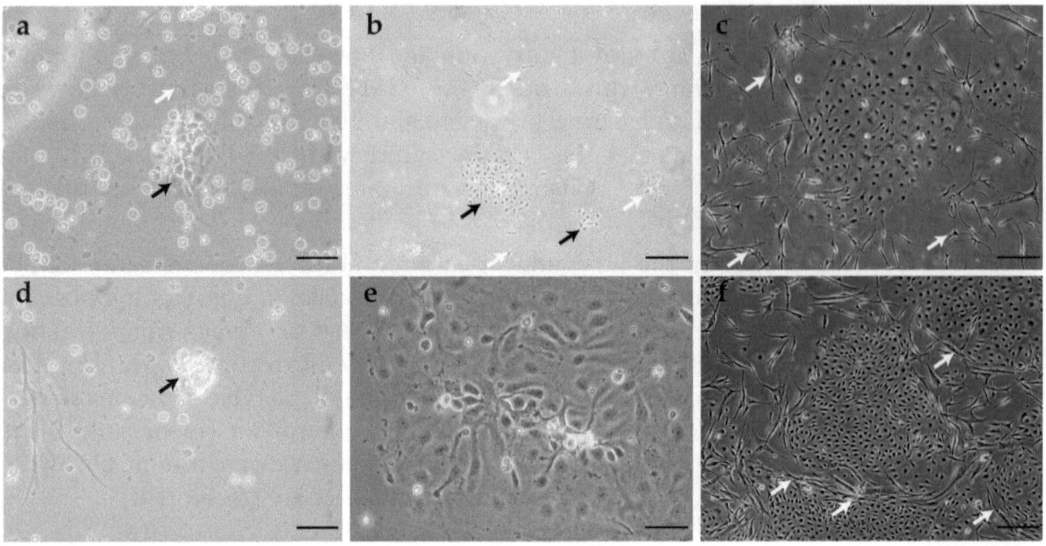

Fig. 3 Formation of endothelial cell colonies from extruded tubules. Migration and proliferation of endothelial cells from the previously extracted tubules after (**a**, **d**) 2 days, (**b**, **e**) 4 days, and (**c**, **f**) 10 days. (**e**) Endothelial cells were extracted from a lipectomy skin sample. (**f**) Endothelial cells were extracted from skin obtained after breast surgery. Black arrows indicate endothelial cells whereas white arrows indicate fibroblasts. (**e**) Many fibroblasts are present, surrounding a small endothelial colony. (**f**) Endothelial cells represent the main population but the colonies are surrounded by fibroblasts. Scale bars: (**a**, **c**, **d**, **f**) 100 μm, (**b**) 250 μm, (**e**) 10 μm

9. Resuspend the cells in HBSS with 5% FBS at the minimum concentration of 2×10^6 cells/mL.
10. Add the amount of previously rinsed beads as determined in **step 3**.
11. Incubate on ice for 30 min under slight agitation.
12. After 30 min, dilute twofold with PBS-0.1% BSA.
13. Place the tube in the magnetic support and incubate for 2 min.
14. Aspirate the supernatant (or pipette it). Be careful not to disturb the beads accumulated along the magnet.
15. Remove the tube from the magnetic support.
16. Add 2 mL of PBS-0.1% BSA.
17. Repeat **steps 13–16** four times.
18. Count the cells and plate them at a minimum of 5000 cells/ cm^2 in EGM2-MV (be careful to use culture dishes coated with gelatin). Cell number can be as low as 20% of the previously estimated endothelial cell number (*see* **Note 11**).
19. Change culture medium three times per week. Be careful to trypsinize the cells before they reach confluency to maintain their proliferation rate.

20. Cells can be frozen at 1–4 × 10^6 cells per mL in freezing medium (we recommend 90% FBS and 10% DMSO, Subheading 2.3, **item 3**).
21. It is recommended to characterize the resulting HMVEC populations according to parameters important for their intended use (*see* **Note 12**).

4 Notes

1. Endothelial cell extraction and efficacy can greatly vary between skin sample types: newborn foreskin, adult mammary skin, or abdominal skin from lipectomy procedures, which is why it is necessary to inspect the sample and note its characteristics such as the presence of hair follicules or nevi, as well as the color of the dermis (pink dermis is highly vascularized). The method described in this chapter can be adapted to pathological skin: at the LOEX center, this protocol was applied to skin from hypertrophic scars as well as from scleroderma patients [7, 8]. Skin surface can be calculated by placing the skin in a Petri dish and drawing its periphery. Cell yield can be evaluated by calculating the number of endothelial cells obtained after immunomagnetic isolation per cm^2. The sample can also be weighted (after adipose tissue removal) to calculate cell yield. Cells can be isolated from a sample as small as 0.5 cm^2.
2. Precautions must be taken to avoid bacterial and fungal contamination since human skin samples may contain many microorganisms. Fresh antibiotics must be added in the solutions and media shortly before use. The sample needs to be thoroughly washed and tools need to be changed as often as necessary to ensure sterility.
3. If the sample is large enough, it might be useful to embed a piece in OCT for cryosectioning and immunostaining. This will allow to assess the original tissue vascularization and compare it with results obtained with the endothelial cells in engineered tissues if necessary.
4. An alternative to the dispase digestion step is the use of thermolysin. Instead of incubating the skin overnight in dispase (Subheading 3.1, **step 14**), it can be incubated with a thermolysin solution: dissolve thermolysin at the concentration of 500 μg/mL in 0.01 M HEPES supplemented with 1 mM $CaCl_2$ at pH 7.45 [9]. The epidermis can then be peeled in the same way as with dispase digestion and further processed similarly. The dermis tends to be stiffer after the thermolysin digestion.

5. Keratinocytes can be isolated from the peeled epidermis. In **step 17**, place the epidermis in a Petri dish containing some trypsin solution (0.05% with 0.01% EDTA). Transfer in a trypsinization unit with trypsin and incubate under agitation at 37 °C for 20 min. To review the complete protocol, *see* [10].
6. Instead of discarding the dermis after extrusion, it can be digested with collagenase H, allowing fibroblast isolation from the same sample [10].
7. Cells may be difficult to count as many of them can be found in tubules. If you need a precise cell count, you can add a step of trypsin digestion (5 min at 37 °C in a 0.05% trypsin–0.01% ethylenediaminetetraacetic acid (EDTA) solution) on the cell pellet to obtain a higher proportion of single cells in the suspension as it detaches the cells from the extracellular matrix.
8. Fibroblasts possess high proliferation rates and can overgrow and displace endothelial cell colonies. To prevent this, the culture needs to be monitored daily before purification. Endothelial cells migrating from the explants will form colonies. These colonies should not be surrounded by confluent fibroblasts or the latter will proliferate and finally displace endothelial cells. If fibroblasts are growing too fast, purification must be carried out before they overgrow the endothelial cells. If applicable, fibroblasts can be carefully removed by aspiration (*see* Subheading 3.2, **step 12**).
9. For the magnetic selection, the ratio of beads per endothelial cells is a critical parameter. This involves estimating the proportion of contaminating fibroblasts. For example, if midsize endothelial colonies are present but surrounded by many fibroblasts, you need to consider that you may have as many fibroblasts as endothelial cells. In this case, add one bead for two cells of the total cell count, as you estimate that only one of the two cells is an endothelial cell. If the bead number is too low, some endothelial cells will be lost. On the contrary, if too many beads are present, they may impede proper cell growth after purification and more fibroblasts can be present. In your estimation, you need to consider that HMVEC are smaller than fibroblasts and grow at a higher density (for example, in a confluent 75 cm^2 flask you can obtain 2–3 million fibroblasts whereas you will obtain 4–6 million endothelial cells).
10. It is of utmost importance to keep cells in cold solutions on ice during the whole purification process. Indeed, fibroblasts tend to intake beads by phagocytosis, which will lead to a mixed population of endothelial cells and fibroblasts, reducing purification efficacy. If needed, the immunoselection process can be repeated after 1–2 passages to increase cell purity.

11. The viability of the isolated endothelial cells can be validated using a marker such as trypan blue or a flow cytometry viability marker.
12. HMVEC can be characterized prior to use by seeding them on gelatin-coated coverslips followed by immunostainings for endothelial cell markers such as CD31, von Willebrand factor, and VE-cadherin. Cells obtained by the method described in this chapter have been used to produce vascularized tissue-engineered skin and 3D constructs presenting blood capillaries which can be modulated by endogenous as well as exogenous growth factors and can recruit pericyte-like cells [11–20]. These cells can also be used for angiogenesis assays [21].

Acknowledgments

The authors acknowledge the precious help and expertise of Sebastien Larochelle, as well as the contribution of Adele Mauroux for the cell isolation. We acknowledge the support of the Centre de recherche du CHU de Québec – Université Laval, of the Fonds de recherche du Québec-Santé (FRQS), and of the Quebec Network for cell and tissue therapies—ThéCell (a thematic network funded by the FRQS).

References

1. Chi J-T, Chang HY, Haraldsen G, Jahnsen FL, Troyanskaya OG, Chang DS et al (2003) Endothelial cell diversity revealed by global expression profiling. Proc Natl Acad Sci 100 (19):10623–10628
2. Thurston G, Daly C (2012) The complex role of angiopoietin-2 in the angiopoietin–tie signaling pathway. Cold Spring Harb Perspect Med 2(9):a006650
3. Kriehuber E, Breiteneder-Geleff S, Groeger M, Soleiman A, Schoppmann SF, Stingl G et al (2001) Isolation and characterization of dermal lymphatic and blood endothelial cells reveal stable and functionally specialized cell lineages. J Exp Med 194(6):797–808
4. Albelda SM (1991) Molecular and cellular properties of PECAM-1 (endoCAM/CD31): a novel vascular cell-cell adhesion molecule. J Cell Biol 114(5):1059–1068
5. Cao G, Fehrenbach ML, Williams JT, Finklestein JM, Zhu J-X, DeLisser HM (2009) Angiogenesis in platelet endothelial cell adhesion molecule-1-null mice. Am J Pathol 175 (2):903–915
6. Richard L, Velasco P, Detmar M (1998) A simple immunomagnetic protocol for the selective isolation and long-term culture of human dermal microvascular endothelial cells. Exp Cell Res 240(1):1–6
7. Bellemare J, Roberge CJ, Bergeron D, Lopez-Vallé CA, Roy M, Moulin VJJ (2005) Epidermis promotes dermal fibrosis: role in the pathogenesis of hypertrophic scars. J Pathol 206 (1):1–8
8. Corriveau MP, Boufaied I, Lessard J, Chabaud S, Senécal JL, Grodzicky T et al (2009) The fibrotic phenotype of systemic sclerosis fibroblasts varies with disease duration and severity of skin involvement: reconstitution of skin fibrosis development using a tissue engineering approach. J Pathol 217(4):534–542
9. Germain L, Rouabhia M, Guignard R, Carrier L, Bouvard V, Auger FA (1993) Improvement of human keratinocyte isolation and culture using thermolysin. Burns 19 (2):99–104
10. Rochon MH, Gauthier MJ, Auger FA, Germain L (2001) Simultaneous isolation of keratinocytes and fibroblasts from a human cutaneous biopsy for the production of

autologous reconstructed skin. Can J Chem Eng 79(4):663–667

11. Moulin VJ, Mayrand D, Laforce-Lavoie A, Larochelle S, Genest H (2011) In vitro culture methods of skin cells for optimal skin reconstruction by tissue engineering. In: Eberli D (ed) Regenerative medicine and tissue engineering—cells and biomaterials. InTech, London. https://doi.org/10.5772/20341
12. Aubin K, Vincent C, Proulx M, Mayrand D, Fradette J (2015) Creating capillary networks within human engineered tissues: Impact of adipocytes and their secretory products. Acta Biomater 11:333–345
13. Mayrand D, Laforce-Lavoie A, Larochelle S, Langlois A, Genest H, Roy M et al (2012) Angiogenic properties of myofibroblasts isolated from normal human skin wounds. Angiogenesis 15(2):199–212
14. Gibot L, Galbraith T, Huot J, Auger FA (2010) A preexisting microvascular network benefits in vivo revascularization of a microvascularized tissue-engineered skin substitute. Tissue Eng Part A 16(10):3199–3206
15. Gibot L, Galbraith T, Huot J, Auger FAA (2013) Development of a tridimensional microvascularized human skin substitute to study melanoma biology. Clin Exp Metastasis 30(1):83–90
16. Berthod F, Symes J, Tremblay N, Medin JA, Auger FAA (2012) Spontaneous fibroblast-derived pericyte recruitment in a human tissue-engineered angiogenesis model in vitro. J Cell Physiol 227(5):2130–2137
17. Tremblay P-LL, Hudon V, Berthod F, Germain L, Auger FAA (2005) Inosculation of tissue-engineered capillaries with the host's vasculature in a reconstructed skin transplanted on mice. Am J Transplant Off J Am Soc Transplant Am Soc Transplant Surg 5 (5):1002–1010
18. Rochon M-HH, Fradette J, Fortin V, Tomasetig F, Roberge CJ, Baker K et al (2010) Normal human epithelial cells regulate the size and morphology of tissue-engineered capillaries. Tissue Eng Part A 16 (5):1457–1468
19. Tremblay P-LL, Berthod F, Germain L, Auger FAA (2005) In vitro evaluation of the angiostatic potential of drugs using an endothelialized tissue-engineered connective tissue. J Pharmacol Exp Ther 315(2):510–516
20. Proulx M, Safoine M, Mayrand D, Aubin K, Maux A, Fradette J (2016) Impact of TNF and IL-1β on capillary networks within engineered human adipose tissues. J Mater Chem B 4 (20):3608–3619
21. Merjaneh M, Langlois A, Larochelle S, Cloutier CB, Ricard-Blum S, Moulin VJJ (2017) Pro-angiogenic capacities of microvesicles produced by skin wound myofibroblasts. Angiogenesis 20(3):385–398

Chapter 8

Isolation of Stromal Vascular Fraction by Fractionation of Adipose Tissue

Joris A. van Dongen, Martin C. Harmsen, and Hieronymus P. Stevens

Abstract

Adipose tissue-derived stromal cells (ASCs) are a promising candidates for cellular therapy in the field of regenerative medicine. ASCs are multipotent mesenchymal stem cell-like and reside in the stromal vascular fraction (SVF) of adipose tissue with the capacity to secrete a plethora of pro-regenerative growth factors. Future applications of ASCs may be restricted through (trans)national governmental policies that do not allow for use of nonhuman-derived (non-autologous) enzymes to isolate ASC. Besides, enzymatic isolation procedures are also time consuming. To overcome this issue, nonenzymatic isolation procedures to isolate ASCs or the SVF are being developed, such as the fractionation of adipose tissue procedure (FAT). This standardized procedure to isolate the stromal vascular fraction can be performed within 10–12 min. The short procedure time allows for intraoperative isolation of 1 mL of stromal vascular fraction derived from 10 mL of centrifuged adipose tissue. The stromal vascular fraction mostly contains blood vessels, extracellular matrix, and ASCs. However, based on the histological stainings an interdonor variation exists which might result in different therapeutic effects. The existing interdonor variations can be addressed by histological stainings and flow cytometry.

Key words Stromal vascular fraction, Adipose tissue, Adipose tissue-derived stromal cells, Fractionation, Regenerative medicine, Cell therapy

1 Introduction

Adipose tissue-derived stromal cells (ASCs) are a promising therapeutic cell type for regenerative purposes because of their ability to differentiate in multiple cell lineages and their ability to secrete a plurality of pro-regenerative growth factors [1, 2]. ASCs are multipotent stem cell-like stromal cells, which are abundantly present in adipose tissue and easily isolated. In adipose tissue, ASCs are attached around vessels as pericytes and supra-adventitial cells in the stromal vascular fraction (SVF) [3, 4]. The SVF of adipose tissue contains all non-adipocyte cells (*e.g.*, immune cells, fibroblasts, endothelial cells, smooth muscle cells, ASCs) [5].

Sophie Böttcher-Haberzeth and Thomas Biedermann (eds.), *Skin Tissue Engineering: Methods and Protocols*, Methods in Molecular Biology, vol. 1993, https://doi.org/10.1007/978-1-4939-9473-1_8,

The therapeutic potential of ASCs is thoroughly investigated clinically for bone and cartilage repair [6], dermal wound healing and fibrosis [7, 8], and myocardial infarction [9, 10], as well as in nonclinical research for tissue engineering purposes like skin tissue [11] or engineered blood vessels [12]. However, the clinical use of ASCs has become a major challenge because the "classical" enzyme-based isolation methods are legally restricted in many countries. Enzymatic isolation methods are time-consuming procedures which require non-autologous materials such as enzymes and animal-derived products [13]. For those reasons, there is an inherent risk of contamination of the isolated ASCs or SVF cells. Moreover, to generate sufficient numbers of ASCs, culturing and expansion of ASCs are needed. The expansion of ASCs for clinical use requires specialized culture labs (Good Manufacturing Practice facilities (cGMP)) which renders the clinical application of ASCs a costly business.

Therefore, nonenzymatic intraoperative isolation procedures to yield a therapeutic cell fraction from adipose tissue are being developed to date [13]. Nonenzymatic, mostly mechanical, isolation procedures are faster and less expensive than enzymatic isolation procedures. Furthermore, nonenzymatic isolation procedures do not require non-autologous biological materials and can therefore be used intraoperatively. Nonenzymatic isolation procedures should not be confused with emulsification procedures that are used to increase the injectability of adipose tissue [13, 14]. In contrast to nonenzymatic isolation procedures, emulsification procedures are not able to disrupt adipocytes. Nonenzymatic isolation procedures often result in a SVF with most of the cell-cell and cell-matrix communications intact (the so-called tissue SVF or tSVF), while enzymatic isolation procedure results in a SVF with only single cells because enzymes disrupt all communication between cells and matrix (the so-called cellular SVF or cSVF) [13]. Clinically, the tSVF and cSVF might have a different therapeutic effect as single cells tend to migrate out of the injection area within the first 24 h after injection [15]. In tSVF, the ASCs are still attached around vessels and embedded in the extracellular matrix, which might result in higher retention rates.

The fractionation of adipose tissue procedure (FAT) yields the tSVF in a nonenzymatic manner [16]. This tSVF is an enrichment of blood vessels, extracellular matrix, as well as ASCs by the disruption of adipocytes. The ASCs isolated from the tSVF are not affected in their function, phenotype, and colony formation capability. Moreover, the high amount of extracellular matrix present in the tSVF may serve as a natural scaffold to deliver and guide cells (*e.g.*, ASCs) in their proliferation as well as differentiation. The extracellular matrix in tSVF contains a large number of vessels as well, which can augment vascularization and perfusion. These latter two are important for appropriate wound healing which is frequently impaired in patients suffering from systemic diseases such as diabetes. Therefore, the isolated tSVF by the FAT procedure might be suitable for skin tissue engineering in vivo to augment (diabetic) wound or ulcer healing.

Several clinical studies have already shown the beneficial influence of adipose tissue or the stromal vascular fraction on dermal wound healing [8, 17–19]. By virtue of the FAT procedure, which is easily standardized, we previously showed that the tSVF composition is subject to interdonor variation. The clinical application of tSVF demands standardized characterization methods. The existing standardized endpoints and methods to validate the isolation procedures and their cellular product are difficult to perform because these methods are time consuming, complex, and expensive. Thus far, no quick intraoperative characterization methods are available. Therefore, we propose easier standardized methods to validate the isolation procedures and their cellular product.

2 Materials

2.1 Liposuction of Adipose Tissue

1. Human subcutaneous liposuction-derived adipose tissue.
2. Scalpel.
3. Modified Klein's solution: 500 mL Saline, 20 mL lidocaine, 2% epinephrine, 2 mL bicarbonate.
4. Sorenson-type lipoharvesting cannula.
5. 50 mL Luer-Lock syringe.

2.2 Fractionation of Adipose Tissue Procedure

1. 10 mL Luer-Lock syringe.
2. Centrifuge with the capability to go to 956 × *g*.
3. Gauge.
4. Fractionator (Luer-to-Luer connector with three holes of 1.4 mm inside).

2.3 Live/Dead Assay of tSVF

1. 0.001% Carboxyfluorescein diacetate succinimidyl ester (CFDA-SE) in serum-free Dulbecco's modified Eagle's medium (DMEM).
2. 0.001% Propium iodide (PI) in serum-free DMEM.
3. 2% Paraformaldehyde (PFA).
4. 4′,6-Diamidino-2-phenylindole (DAPI).

2.4 Histological Characterization of tSVF

1. 10% Formalin.
2. 1.8% Agarose solution.
3. Xylol.
4. 100%, 96%, 70% alcohol.

5. 0.1 M Tris/hydrochloric acid (HCl) buffer pH 9.0.
6. 10 mM Tris/1 mM ethylenediaminetetraacetic acid (EDTA) buffer pH 9.0.
7. 3% Hydrogen peroxidase/phosphate-buffered saline (PBS).
8. α-Smooth muscle actin (SMA, Abcam #ab7871, Cambridge, UK): 1:200, 1% BSA, 1% human serum in PBS.
9. Perilipin A (Abcam #ab3526, Cambridge, UK): 1:200, 1% BSA, 1% human serum in PBS.
10. von Willebrand factor (vWF, Dako #A0082, Glostrup, Denmark): 1:200, 1% BSA, 1% swine serum in PBS.
11. 1% Bovine serum albumin (BSA).
12. 1% Human serum.
13. 1% Swine serum.
14. Polyclonal Rabbit anti-Mouse (Dako #P0260, Glostrup, Denmark): 1:100, 1% BSA, 1% human serum in PBS for α-SMA.
15. Polyclonal Goat anti-Rabbit (Dako #P0448, Glostrup, Denmark): 1:100, 1% BSA, 1% human serum in PBS for perilipin A.
16. Polyclonal Swine anti-Rabbit (Dako #P0217, Glostrup, Denmark): 1:100, 1% BSA, 1% human serum in PBS for vWF.
17. 3,3′-Diaminobenzidine (DAB, Sigma Life Science, St. Louis, MO).
18. Hematoxylin.
19. Mounting solution.

2.5 Masson's Trichrome Staining of tSVF

1. 4% Paraformaldehyde (PFA).
2. 1% Triton X-100.
3. Bouin fixative: 36 mL Picric acid (saturated), 12 mL 37% formalin, 2 mL acetic acid (glacial).
4. Weigert's iron hematoxylin (25 mL of stock solution A and 25 mL of stock solution B): Stock solution A: 0.5 g hematoxylin, 50 mL 100% alcohol.

   Stock solution B: 0.6 g iron chloride, 49.5 mL demiwater, 0.5 mL hydrochloric acid.
5. Biebrich scarlet-acid fuchsin: 0.5 g Biebrich scarlet, 0.5 g acid fuchsin, 49.5 mL demiwater, 0.5 mL acetic acid (glacial).
6. Phosphomolybdic-phosphotungstic acid: 2.5 g Phosphomolybdic acid, 2.5 g phosphotungstic acid, 50 mL demiwater.
7. Aniline blue: 1.3 g Aniline blue, 1 mL acetic acid (glacial), 49 mL demiwater.
8. 1% Acetic acid: 0.5 mL Acetic acid (glacial), 49.5 mL demiwater.

2.6 Enzymatic Preparation of Cellular SVF (cSVF) Derived from tSVF

1. 0.1% Bacterial collagenase A solution: 50 mg Bacterial collagenase A, 50 mL PBS, 1% BSA.
2. 100 μm Filters.
3. Lymphoprep.
4. Lysis buffer.
5. FACS buffer (PBS/0.5% BSA).

3 Methods

3.1 Liposuction of Adipose Tissue

1. Make a small stab incision of 1 cm on the donor site (preferably legs or abdomen).
2. Infiltrate the donor site with a modified Klein's solution.
3. Place a Sorenson-type lipoharvesting cannula into the stab incision.
4. Connect a 50 mL Luer-Lock syringe to a Sorenson-type harvesting cannula.
5. Pull the plunger backwards to allow for negative pressure and use a surgical clamp or syringe snap lock to maintain the negative pressure.
6. Start harvesting by moving the harvesting cannula forward and backward.
7. Replace the full 50 mL syringe by an empty 50 mL syringe and start again from **step 4**.

3.2 Fractionation of Adipose Tissue Procedure

1. Divide the harvested adipose tissue in 10 mL Luer-Lock syringes without plunger.
2. Decant the adipose tissue for 5 min at room temperature.
3. Remove infiltration fluid by opening the cap of the 10 mL syringe.
4. Refill syringe to 10 mL and centrifuge the syringe without plunger at 3000 rpm with a 9.5 cm radius fixed-angle rotor or at 960 × *g* for 2.5 min at room temperature.
5. Remove infiltration fluid by opening the cap of the 10 mL syringe (*see* **Note 1**).
6. Remove oil (disrupted adipocytes) by turning the syringe upside down and prevent the adipose tissue from leaking with the use of a gauge.
7. Refill syringe to 10 mL of centrifuged adipose tissue and place the plunger back.
8. Connect the 10 mL syringe with centrifuged adipose tissue to the fractionator and connect an empty 10 mL syringe with plunger to the other side of the fractionator.

9. Push the adipose tissue extensively forward and backward 30 times (*see* **Note 2**).
10. Centrifuge the syringe without plunger at 3000 rpm with a 9.5 cm radius fixed-angle rotor or at 960 × *g* at room temperature for 2.5 min.
11. Remove infiltration fluid by opening the cap of the 10 mL syringe.
12. Remove oil (disrupted adipocytes) by turning the syringe upside down and prevent the stromal vascular fraction (SVF) from leaking with the use of a gauge.

3.3 Live/Dead Assay of SVF

1. SVF is mixed with preheated (37 °C) 0.001% CFDA-SE and 0.001% (PI) in serum-free DMEM and allow for 30 min of incubation under normal culture conditions (37 °C).
2. Wash the SVF with PBS three times.
3. Fix the SVF with 2% PFA for 30 min.
4. Wash the SVF with PBS three times.
5. Stain the nuclei DAPI in the dark for 30 min.
6. Wash the SVF with PBS three times.

3.4 Histological Characterization of SVF

1. Fix the isolated SVF in 10% formalin overnight at 4 °C.
2. Embed the SVF in a preheated (60 °C) 1.8% agarose solution (1:2).
3. Place the SVF/agarose solution at 4 °C for 30 min.
4. Dehydrate samples with the following steps in sequence at room temperature:
   (a) 50% Alcohol for 30 min.
   (b) 70% Alcohol for 30 min.
   (c) 96% Alcohol for 30 min.
   (d) Twice in 100% alcohol for 30 min.
   (e) Twice in xylol for 30 min.
5. Embed the samples in paraffin.
6. Cut 4 mm sections and deparaffinize them in xylol for 15 min.
7. Refresh the xylol and place the samples for another 10 min in xylol.
8. Move the samples and place them in 100% alcohol for 10 min, then in 96% alcohol for 3 min, and finally in 70% alcohol for 3 min at room temperature.
9. Wash the samples in demiwater for 3 min.
10. Incubate samples overnight with 0.1 M Tris/HCl buffer (pH 9.0) at 80 °C for *α*-SMA as well as for perilipin A staining

and with 10 mM Tris/1 mM EDTA buffer (pH 9.0) at 80 °C for vWF.

11. Cool down to room temperature for 30 min.
12. Wash samples with PBS three times.
13. Endogenous peroxidase activity is blocked with 3% hydrogen peroxidase in PBS at room temperature for 30 min.
14. Wash samples with PBS three times and incubate samples with primary antibodies at room temperature for 60 min: for α-SMA, for perilipin A, and for vWF.
15. Wash samples with PBS three times.
16. Incubate samples with secondary antibodies for 30 min at room temperature: polyclonal rabbit anti-mouse for α-SMA, polyclonal goat anti-rabbit for perilipin A, and polyclonal swine anti-rabbit for vWF.
17. Wash samples with PBS three times.
18. Incubate α-SMA sample with a third antibody at room temperature for 30 min with polyclonal swine anti-rabbit.
19. Wash α-SMA samples with PBS three times.
20. Incubate all samples with DAB for 10 min at room temperature in the dark.
21. Wash all samples in demiwater three times for 5 min.
22. Incubate all samples with hematoxylin for 1.5 min.
23. Rinse samples in water for 4 min.
24. Mount all the samples and place a coverslip (*see* **Note 3**).

3.5 Masson's Trichrome Staining of tSVF

1. Follow **steps 1** till **9** of Subheading 3.3.
2. Fix samples in 4% PFA for 60 min.
3. Permeabilize samples in 1% Triton X-100 for 10 min.
4. Wash samples with PBS three times.
5. Fix samples in Bouin fixative at 51 °C for 5 min.
6. Wash samples in demiwater until color disappears.
7. Incubate samples with Weigert's iron hematoxylin for 20 min.
8. Wash samples in demiwater until color disappears.
9. Incubate samples with Biebrich scarlet-acid fuchsin for 20 min.
10. Wash samples in demiwater for 2 min.
11. Incubate samples in phosphomolybdic-phosphotungstic acid for 12 min.
12. Wash samples in demiwater until color disappears.
13. Incubate samples in aniline blue for 7 min.
14. Wash samples in demiwater for 2 min.

15. Incubate samples in 1% acetic acid for 5 s.
16. Wash samples in demiwater until color disappears.
17. Dry samples for 20 min.
18. Mount samples in Permount.

3.6 Enzymatic Preparation of Cellular SVF (cSVF) Derived from tSVF

1. Follow **steps 1** till **12** of Subheading 3.2.
2. Wash the isolated tSVF with PBS three times.
3. Add 0.1% collagenase A solution 1:1 with tSVF.
4. Stir the collagenase/tSVF mixture in a water bath at 37 °C for 1.5 h.
5. Centrifuge the sample at 600 × *g* at room temperature for 10 min.
6. Remove supernatant.
7. Collect cell pellet in PBS/1% BSA.
8. Filter the collagenase/tSVF mixture through filters.
9. Centrifuge the sample at 600 × *g* at room temperature for 10 min.
10. Repeat **steps 6–8**.
11. Remove supernatant.
12. Collect cell pellet in 30 mL of PBS/1% BSA.
13. Put 15 mL of lymphoprep in a 50 mL tube.
14. Gently add the 30 mL of PBS/1% BSA with the sample.
15. Centrifuge the sample at 1000 × *g* at 4 °C for 20 min and put the brake on 0.
16. Remove the upper layer.
17. Take cells from the interphase carefully.
18. Add PBS/1% BSA to the cells.
19. Centrifuge the sample at 800 × *g* at 8 °C for 10 min.
20. Remove supernatant.
21. Resuspend cell pellet in lysis buffer and place the sample on ice for 5 min.
22. Centrifuge the sample at 800 × *g* at 8 °C for 10 min.
23. Remove supernatant.
24. Repeat **steps 19** and **21** till all erythrocytes are disrupted and the red color has disappeared.
25. Resuspend cell pellet in FACS buffer and divide cells in multiple tubes. The number of tubes depends on the number of subset of CD markers used. Additionally, one tube will function as a blank control and for each fluorophore used an extra

Table 1
The phenotype of the most important adipose-derived ($CD45^{neg}$) cell types based on CD marker expression in freshly isolated tSVF

Cell type	CD31	CD34	CD45	CD90	CD105	CD146
Progenitor pericyte	Negative	Positive	Negative	Positive	Negative	Positive
Pericyte	Negative	Negative	Negative	Positive	Negative	Positive
Supra-adventitial cell	Negative	Positive	Negative	Positive	Negative	Negative
Fibroblast	Negative	Negative	Negative	Positive	Negative	Negative
Adipose-derived stromal cell	Negative	Positive	Negative	Positive	Low percentage positive	Negative
Vascular endothelial cell	Positive	Positive	Negative	Positive	Low percentage positive	Positive
Endothelial cell	Low percentage positive	Positive	Negative	Positive	Low percentage positive	Positive

tube will be used for the fluorophore-specific isotype control (*see* **Note 4**).

26. Centrifuge cells at 300 × *g* at 4 °C for 5 min.
27. Resuspend cell pellet in 100 μL FACS buffer.
28. Keep one tube of cells unlabeled and put on ice in the dark for 30 min. This tube functions as a blank control to set up the flow cytometer.
29. Incubate a tube of cells on ice with 5 μL of the preferred fluorophore-conjugated antibodies (1:20) in the dark for 30 min (Table 1 and *see* **Note 5**).
30. Incubate the other tubes of cells on ice with 5 μL of the specific fluorophore isotype control (1:20) in the dark for 30 min.
31. Wash samples with 2 mL FACS buffer.
32. Centrifuge cells at 300 × *g* at 4 °C for 5 min.
33. Remove supernatant.
34. Repeat **steps 10** till **12** three times.
35. Resuspend cell pellet in 300 μL FACS buffer.
36. Proceed to FACS for CD marker analysis.

4 Notes

1. The fractionation of adipose tissue procedure only works when the harvested adipose tissue is separated from all the infiltration fluid and oil by centrifugation at a high speed (960 × *g*). When small amounts of infiltration fluid are left behind, none of the

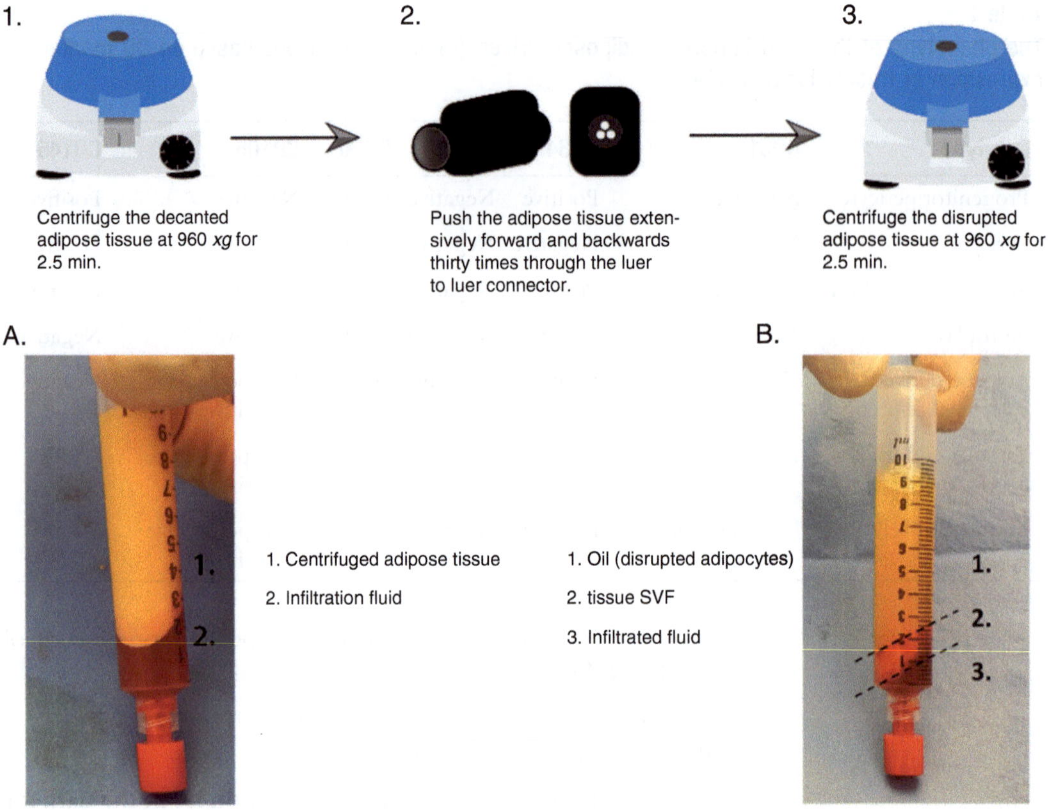

Fig. 1 The fractionation of adipose tissue procedure (the FAT procedure). (**a**) The composition of one-time-centrifuged adipose tissue (1. adipose tissue, 2. infiltration fluid) at 960 × *g* for 2.5 min. In some cases, oil already appears after the first round of centrifugation. (**b**) The composition of adipose tissue (1. oil (disrupted adipocytes), 2. tissue SVF, 3. infiltration fluid) after centrifugation at 960 × *g* and disruption by means of the fractionator and centrifugation at 960 × *g*. SVF = stromal vascular fraction

adipocytes will be disrupted and therefore no oil will appear after the final centrifugation step. In case if fibrotic adipose tissue is processed by the fractionation of adipose tissue procedure, the fractionator can clog. The fractionator can be cleaned manually with 100% ethanol (Fig. 1).

2. In case some infiltration fluid is left behind, the harvested adipose tissue will turn white when the harvested adipose tissue is pushed forward and backward through the Luer-to-Luer connector. After the second round of centrifugation, no oil will appear and therefore the stromal vascular fraction cannot be isolated. This is called emulsified adipose tissue; the liquid content is mixed with the adipose tissue content and allows for a better injectable fraction (Fig. 2).

3. Results of the immunohistochemistry images can be analyzed with the use of ImageJ software (freeware, NIH). α-SMA and vWF stainings are measured by drawing a line around the tissue

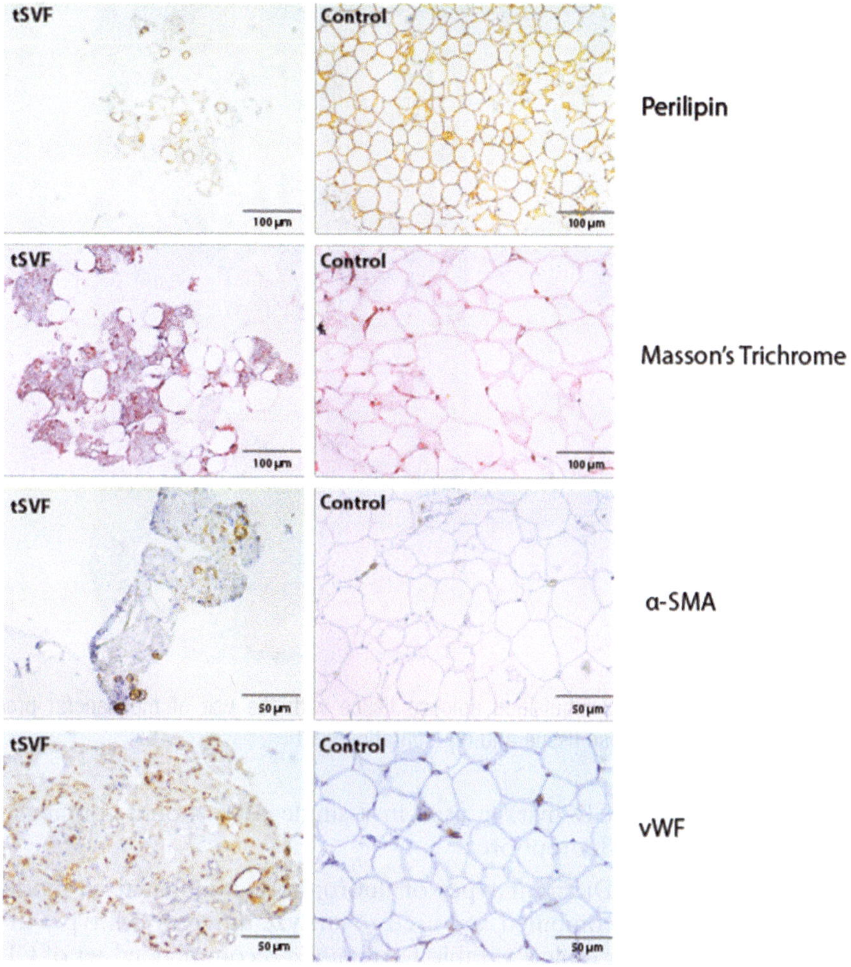

Fig. 2 Immunohistochemistry staining examples of how tSVF and unprocessed adipose tissue (control) should look like when a perilipin, Masson's trichrome, α-smooth muscle actin (α-SMA), and von Willebrand factor (vWF) staining are performed

sample of interest and set a threshold to separate positive cells from negative cells. Perilipin A staining can be analyzed by manual counting of the positive adipocytes (Fig. 3).

4. For a complete characterization of the tSVF, flow cytometry analysis of enzymatic isolated tSVF is advised. Each tube contains all the CD markers used to analyze the preferred cell type present in the tSVF (regardless of a positive or negative expression of the CD marker on the surface of the cell type). It is possible to analyze multiple cell types with the same CD markers in one tube. The maximum number of CD markers used in a single tube depends on the chosen fluorophore (*e.g.*, allophycocyanin (APC) and fluorescein isothiocyanate (FITC)). Each

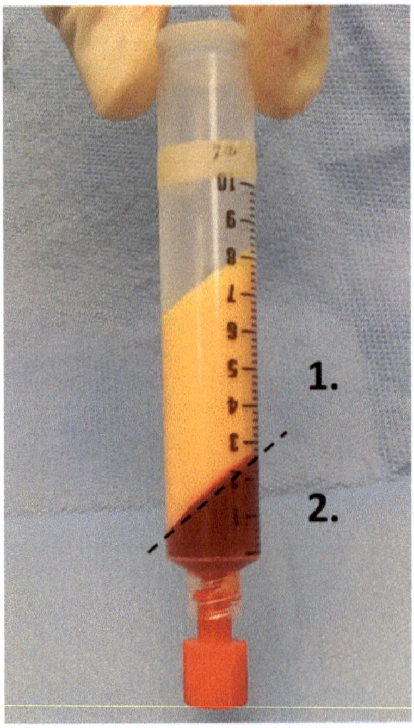

Fig. 3 Emulsified adipose tissue with the use of the nanofat procedure. (**1**) Adipose tissue and (**2**) infiltration fluid

CD marker used in a single tube should contain a different fluorophore.

5. Different types of fluorophore-conjugated antibodies can be combined and used to analyze different cell types in the tSVF (Table 1). Table 1 contains a recommended set of CD markers to analyze the most important adipose-derived cell types in the tSVF [13]. However, there is no consensus regarding the correct subset of CD markers for each cell type [3, 4, 13, 20–22].

References

1. Zuk PA, Zhu M, Mizuno H et al (2001) Multilineage cells from human adipose tissue: implications for cell-based therapies. Tissue Eng 7 (2):211–228
2. Roux S, Bodivit G, Bartis W et al (2015) In vitro characterization of patches of human mesenchymal stromal cells. Tissue Eng A 21 (3-4):417–425
3. Lin G, Garcia M, Ning H et al (2008) Defining stem and progenitor cells within adipose tissue. Stem Cells Dev 17(6):1053–1063
4. Corselli M, Chen CW, Sun B, Yap S, Rubin JP, Peault B (2012) The tunica adventitia of human arteries and veins as a source of mesenchymal stem cells. Stem Cells Dev 21 (8):1299–1308
5. Bourin P, Bunnell BA, Casteilla L et al (2013) Stromal cells from the adipose tissue-derived stromal vascular fraction and culture expanded adipose tissue-derived stromal/stem cells: a joint statement of the International Federation for Adipose Therapeutics and Science (IFATS) and the International Society for Cellular Therapy (ISCT). Cytotherapy 15(6):641–648
6. Grayson WL, Bunnell BA, Martin E, Frazier T, Hung BP, Gimble JM (2015) Stromal cells and stem cells in clinical bone regeneration. Nat Rev Endocrinol 11(3):140–150

7. Spiekman M, van Dongen JA, Willemsen JC, Hoppe DL, van der Lei B, Harmsen MC (2017) The power of fat and its adipose-derived stromal cells: emerging concepts for fibrotic scar treatment. J Tissue Eng Regen Med 11(11):3220–3235
8. Bura A, Planat-Benard V, Bourin P et al (2014) Phase I trial: the use of autologous cultured adipose-derived stroma/stem cells to treat patients with non-revascularizable critical limb ischemia. Cytotherapy 16(2):245–257
9. Jiang Y, Chang P, Pei Y et al (2014) Intramyocardial injection of hypoxia-preconditioned adipose-derived stromal cells treats acute myocardial infarction: an in vivo study in swine. Cell Tissue Res 358(2):417–432
10. Chen L, Qin F, Ge M, Shu Q, Xu J (2014) Application of adipose-derived stem cells in heart disease. J Cardiovasc Transl Res 7 (7):651–663
11. Klar AS, Zimoch J, Biedermann T (2017) Skin tissue engineering: application of adipose-derived stem cells. Biomed Res Int 2017:9747010
12. Parvizi M, Bolhuis-Versteeg LA, Poot AA, Harmsen MC (2016) Efficient generation of smooth muscle cells from adipose-derived stromal cells by 3D mechanical stimulation can substitute the use of growth factors in vascular tissue engineering. Biotechnol J 11 (7):932–944
13. van Dongen JA, Tuin AJ, Spiekman M, Jansma J, van der Lei B, Harmsen MC (2018) Comparison of intraoperative procedures for isolation of clinical grade stromal vascular fraction for regenerative purposes: a systematic review. J Tissue Eng Regen Med 12(1): e261–e274. https://doi.org/10.1002/term.2407
14. Tonnard P, Verpaele A, Peeters G, Hamdi M, Cornelissen M, Declercq H (2013) Nanofat grafting: basic research and clinical applications. Plast Reconstr Surg 132(4):1017–1026
15. Parvizi M, Harmsen MC (2015) Therapeutic prospect of adipose-derived stromal cells for the treatment of abdominal aortic aneurysm. Stem Cells Dev 24(13):1493–1505
16. van Dongen JA, Stevens HP, Parvizi M, van der Lei B, Harmsen MC (2016) The fractionation of adipose tissue procedure to obtain stromal vascular fractions for regenerative purposes. Wound Repair Regen 24(6):994–1003
17. Stasch T, Hoehne J, Huynh T, De Baerdemaeker R, Grandel S, Herold C (2015) Debridement and autologous lipotransfer for chronic ulceration of the diabetic foot and lower limb improves wound healing. Plast Reconstr Surg 136(6):1357–1366
18. Raposio E, Bertozzi N, Bonomini S et al (2016) Adipose-derived stem cells added to platelet-rich plasma for chronic skin ulcer therapy. Wounds 28(4):126–131
19. Conde-Green A, Marano AA, Lee ES et al (2016) Fat grafting and adipose-derived regenerative cells in burn wound healing and scarring: a systematic review of the literature. Plast Reconstr Surg 137(1):302–312
20. Corselli M, Crisan M, Murray IR et al (2013) Identification of perivascular mesenchymal stromal/stem cells by flow cytometry. Cytometry A 83(8):714–720
21. Zimmerlin L, Donnenberg VS, Pfeifer ME et al (2010) Stromal vascular progenitors in adult human adipose tissue. Cytometry A 77 (1):22–30
22. Traktuev DO, Merfeld-Clauss S, Li J et al (2008) A population of multipotent CD34-positive adipose stromal cells share pericyte and mesenchymal surface markers, reside in a periendothelial location, and stabilize endothelial networks. Circ Res 102(1):77–85

Part II

Tissue Engineering Skin and Engineered Skin Models

Chapter 9

Engineering a Multilayered Skin Equivalent: The Importance of Endogenous Extracellular Matrix Maturation to Provide Robustness and Reproducibility

Lydia Costello, Nicola Fullard, Mathilde Roger, Steven Bradbury, Teresa Dicolandrea, Robert Isfort, Charles Bascom, and Stefan Przyborski

Abstract

Human skin equivalents (HSEs) are a valuable tool for both academic and industrial laboratories to further the understanding of skin physiology and associated diseases. Over the last few decades, there have been many advances in the development of HSEs that successfully recapitulate the structure of human skin in vitro; however a main limitation is variability due to the use of complex protocols and exogenous extracellular matrix (ECM) proteins. We have developed a robust and unique full-thickness skin equivalent that is highly reproducible due to the use of a consistent scaffold, commercially available cells, and defined low-serum media. The Alvetex® scaffold technology allows fibroblasts to produce their own endogenous ECM proteins within the scaffold, which alleviates the need for exogenous collagen, and supports the differentiation and stratification of the epidermis. Our full-thickness skin equivalent is generated using a detailed step-by-step protocol, which sequentially forms the multilayered structure of human skin in vitro. This model can be adapted for many downstream applications such as disease modeling and testing of active compounds for cosmetics.

Key words Skin equivalent, Organotypic culture, Alvetex® scaffold, Skin tissue engineering, Endogenous extracellular matrix deposition

1 Introduction

1.1 Structure of Human Skin

Human skin has a complex structure, composed of three morphologically and functionally distinct layers: the epidermis, dermis, and hypodermis. The epidermis acts as a multifunctional physical barrier to withstand mechanical stress, protect against percutaneous pathogen invasion, and prevent unregulated transepidermal water and electrolyte loss [1]. The epidermal compartment has a stratified, squamous organization with a multitude of cell types including mechanosensory Merkel cells, immunoregulatory Langerhans cells, photoprotective melanocytes, and keratinocytes, which

Sophie Böttcher-Haberzeth and Thomas Biedermann (eds.), *Skin Tissue Engineering: Methods and Protocols*, Methods in Molecular Biology, vol. 1993, https://doi.org/10.1007/978-1-4939-9473-1_9,

predominantly constitute 95% of the epidermis [2–4]. Proliferative keratinocytes arise from asymmetrically dividing stem cell populations in the *stratum basale* and they undergo profound morphological and cytoarchitectural changes during programmed sequential differentiation to form the suprabasal *stratum spinosum* and *stratum granulosum* layers [5]. During terminal differentiation, keratinocytes undergo cornification into anuclear, flattened corneocytes that become embedded in the lipid-rich, proteinaceous *stratum corneum*. This apical layer is produced by the covalent cross-linking of synthesized proteins and lipids by calcium-dependent transglutaminases, to produce a water-resistant, protective barrier [6] (Fig. 1).

The epidermis is tightly adhered to the dermis via the interdigitating dermoepidermal junction, which facilitates the exchange of substances and aids polarity of basal keratinocytes [7]. The dermis provides the skin with mechanical strength and elastic recoil,

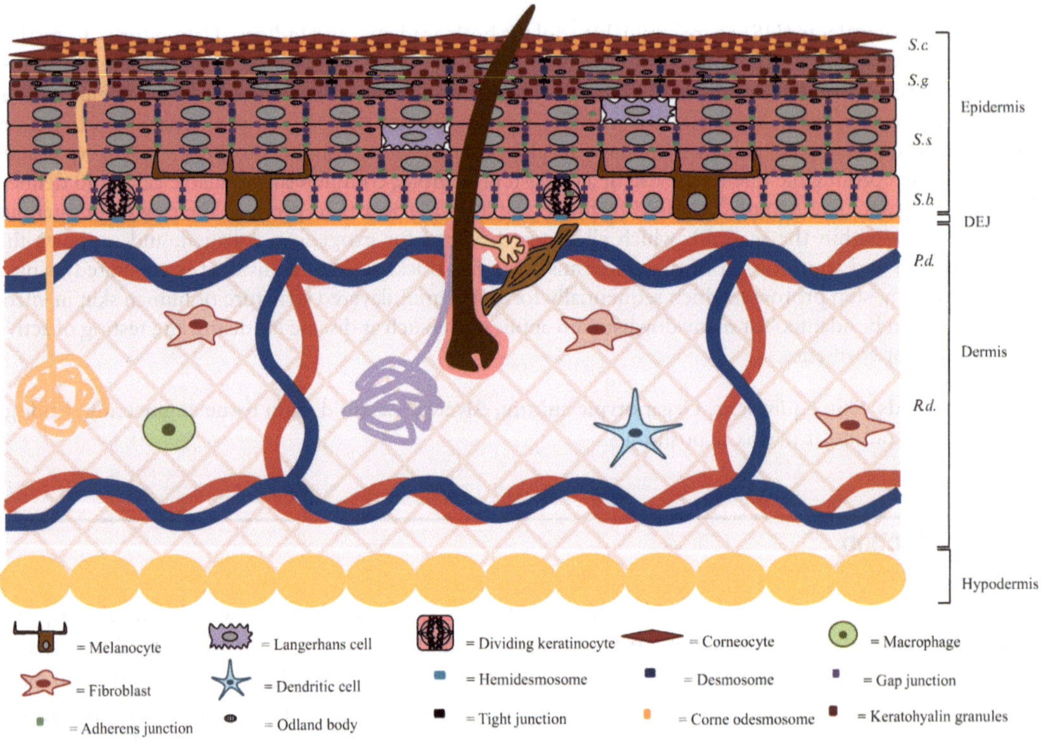

Fig. 1 Schematic of human skin. The skin contains three morphologically distinct layers: the epidermis, dermis, and hypodermis. The epidermis is a stratified epithelial compartment composed of the *stratum basale (S.b.)*, *stratum spinosum* (*S.s.*), *stratum granulosum (S.g.)*, and *stratum corneum* (*S.c.*). The epidermis is supported by the dermis, which is comprised of the papillary (*P.d.*) and reticular (*R.d.*) layers. The dermis is perfused with blood vessels, and contains a multitude of cell types such as fibroblasts, which synthesize the extracellular matrix, and immune cells such as macrophages and dendritic cells. Dermal appendages include hair follicles, sebaceous glands, and sweat glands. The hypodermis is composed of subcutaneous adipose tissue, which provides insulation

and is composed of two morphologically distinct layers: the superficial papillary dermis containing sparsely arranged collagen and elaunin fibers, and the deep reticular dermis containing abundant collagen fibers and a dense elastic network [8]. These networks are embedded in an amorphous, non-fibrous matrix composed of proteoglycans and glycosaminoglycans such as hyaluronic acid, chondroitin sulfate, and dermatan sulfate, which are synthesized by fibroblasts within the dermis (Fig. 1). The hypodermis is a layer of subcutaneous fat that provides insulation and storage.

1.2 Bioengineered Skin Models Bridge the Gap Between Two-Dimensional Tissue Culture and Animal Models

Traditionally, skin research has been conducted by culturing two-dimensional homogeneous keratinocyte or fibroblast populations on plastic substrates. Although this method is widely accepted and allows the use of high-throughput, cost-effective assays, it fails to represent the multilayered complexity of human skin. To more accurately study the in vivo microenvironment, mouse models are also often used. Animal models are favorable for research as they allow the systemic effects of an active compound to be observed, which often cannot be captured with in vitro assays. Despite the biological similarities, there are fundamental architectural discrepancies between murine and human skin such as hair follicle distribution, lack of rête ridges, decreased epidermal thickness, and presence of the panniculus carnosus in mouse skin, which is absent in humans [9]. It was also recently found that only 30% of human skin-associated genes are expressed in murine skin [10]. In addition to these anatomical differences, European Union legislation advocates the replacement, reduction, and refinement of animal use in scientific research, and a complete prohibition on the use of animals for testing active compounds for cosmetics was recently introduced [11].

Organotypic HSEs have been developed which overcome the limitations of in vivo animal models and in vitro two-dimensional cell culture, and provide a valuable platform technology for skin research. Over the last few decades, significant progress has been made in the development of physiologically relevant HSEs that successfully recapitulate the multilayered complexity of human skin. HSEs are used globally in both academic and industrial research laboratories, and they have multifaceted applications in vitro including permeability testing, disease modeling, and screening of active compounds for cosmetic and pharmaceutical development.

1.3 Engineering the Multilayered Structure of Human Skin In Vitro

Engineering HSEs in vitro is technically challenging due to their multilayered, multicellular structure. As a consequence, detailed step-by-step protocols are required. Firstly, fibroblasts are seeded into the chosen matrix: de-epidermized dermis (DED), hydrogel, or scaffold to form the dermal compartment. The epidermis is more complex due to the presence of sequentially differentiated layers.

The accepted method of generating the epidermis involves initially seeding keratinocytes onto the dermis in submerged culture, and culturing in low-calcium conditions to promote proliferation and formation of the *stratum basale*. To achieve epidermal differentiation, the HSEs are then raised to the air-liquid interface in high-calcium conditions, known as the calcium switch. The calcium switch mediates the characteristic basal-to-suprabasal switch of keratin 5/14 to keratin 1/10 [12]. Due to their highly complex structure, the HSEs must be extensively analyzed to prove that they successfully recreate the skin architecture in vitro. Once the HSE has been characterized and validated, it can be used for in vivo and in vitro applications.

To recapitulate the dermal compartment in vitro, many matrices can be used. Human DEDs are acellular dermal substrates that accurately retain the native structure of the ECM. An advantage of using DEDs is that the papillary and reticular layers remain intact, which is technically challenging to achieve with other matrices. As DEDs are donor derived, drawbacks include limited availability and individual differences such as DED thickness, anatomical site, age-dependent differences, and variation in ECM composition and growth factors. DEDs have been shown to produce full-thickness skin equivalents when seeded with keratinocytes, which form a well-differentiated epidermis [13]; however their inherent heterogeneity and lack of reproducibility limit their use in in vitro assays.

To overcome these limitations, full-thickness skin equivalents have been generated using a variety of matrices such as hydrogels and porous scaffolds to mimic the human dermis. Hydrogels resemble the structure of the dermal ECM, as they are usually composed of natural macromolecules such as collagen. Previous studies have shown that they provide a good support for epidermal differentiation and stratification, and the formation of a basement membrane [14]. Limitations of hydrogel use include batch-to-batch variation, the presence of only one extracellular matrix component, and their often animal-derived origin such as bovine or rat. Nonetheless, hydrogels are widely used in organotypic culture due to the rapid generation of the dermal compartment, ability to support an epidermis without keratinocyte infiltration into the dermis, and availability of human collagen for fully humanized skin models [15].

Alternatives to hydrogel matrices are self-assembled dermal compartments, whereby fibroblasts are seeded within the porous scaffolds and stimulated to proliferate and secrete endogenous dermal ECM that is physiologically representative of normal human skin. An example of a commercially available polystyrene scaffold is Alvetex®, which has a thickness of 200 μm and over 90% porosity, to allow the efficient exchange of nutrients, gases, and waste products across a short distance [16] (Fig. 2). The use of a

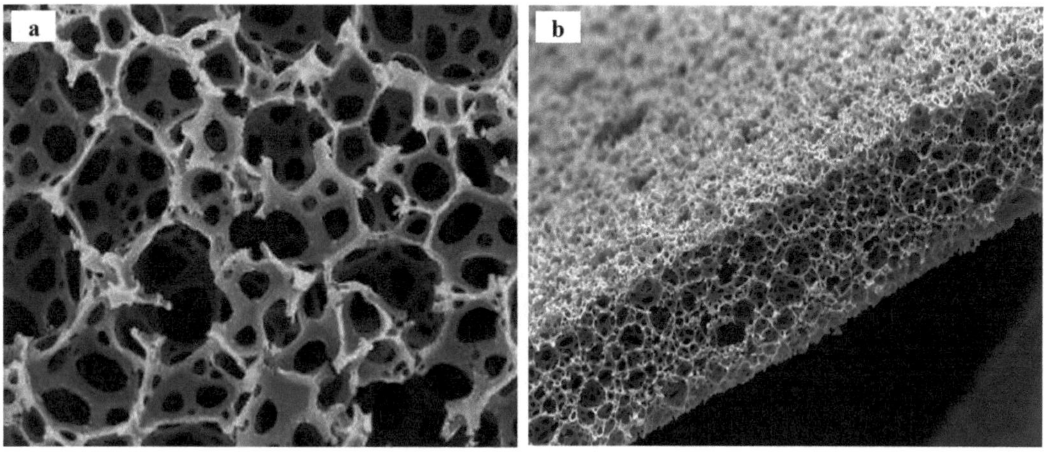

Fig. 2 Alvetex® scaffold. Scanning electron microscopy images of the Alvetex® scaffold showing (**a**) a magnified image of the 30–40 μm pores and (**b**) an overview of the porous scaffold disk. Alvetex® is a polystyrene scaffold that allows cells to be cultured in three dimensions, to more accurately recapitulate an in vivo environment. Images courtesy of Reprocell Europe

scaffold to generate HSEs is favorable as it removes variability, and as the fibroblasts generate their own ECM, it allows changes in dermal ECM composition to be observed, for example in skin aging research or in response to an active compound. A limitation of using scaffold technology is potential keratinocyte infiltration into the dermis; however this can be resolved by optimizing the culture length of the dermal component to allow sufficient production of ECM proteins to support the epidermis.

1.4 Development of Human Skin Equivalents

HSEs have multifaceted applications in dermatological research, and there are many market-leading, commercially available HSEs including EpiDermFT™ (MatTek Corporation, MA, USA), T-skin™ (Episkin, L'Oréal, France), and the Phenion® Full Thickness Skin Model (AG&Co, KGaA, Dusseldorf, Germany). These models are widely used as they are standardized, quality controlled and they resemble the structure of human skin, however as they are pre-made, they lack the flexibility to be tailored for specific downstream applications for example in the generation of aging or diseased models.

HSEs have been developed in academic laboratories to overcome this problem; however many lack reproducibility due to the use of serum and exogenous extracellular matrix proteins [14]. While donor-derived cells have great potential for generating donor-matched skin equivalents for modeling diseased states or personalized medicine [17], they possess intrinsic donor-to-donor heterogeneity, which reduces the reproducibility of an in vitro skin equivalent and is problematic for high-throughput testing in industry. To overcome this problem, two research groups have successfully developed skin equivalents using TERT-immortalized and commercially available cells to increase reproducibility; however further limitations include the use of animal-derived hydrogels with batch-to-batch variability [18, 19].

1.5 The Development of a Robust and Reproducible Full-Thickness Skin Equivalent

The unique skin equivalent developed using Alvetex® technology can overcome the limitations of current HSEs as it is robust and reproducible due to the use of a consistent scaffold, commercially available cells, and defined low-serum media. In addition, fibroblasts make their own endogenous extracellular matrix within the Alvetex® scaffold, which further reduces variability as exogenous animal-derived collagen hydrogels are not required. Unlike commercially available models, it also provides the flexibility to be custom-made for a wide range of downstream applications.

Histological analysis of the full-thickness skin equivalent resembles the structure of human skin. Fibroblasts populate the Alvetex® scaffold, and generate sufficient ECM proteins to support the epidermal compartment. The proliferative keratinocytes within the *stratum basale* are one cell thick, and highly organized with a columnar morphology. The basal keratinocytes undergo sequential differentiation and stratification to form the polyhedral keratinocytes of the *stratum spinosum*, and the characteristic elongated keratinocyte morphology of the *stratum granulosum*. Finally, the keratinocytes undergo terminal differentiation into anuclear, flattened corneocytes within the *stratum corneum* layers, similar to in vivo skin (Fig. 3).

Histological analysis of the dermal compartment shows that fibroblasts are evenly distributed throughout the Alvetex® scaffold (Fig. 4). Unlike many alternative matrices, culturing fibroblasts

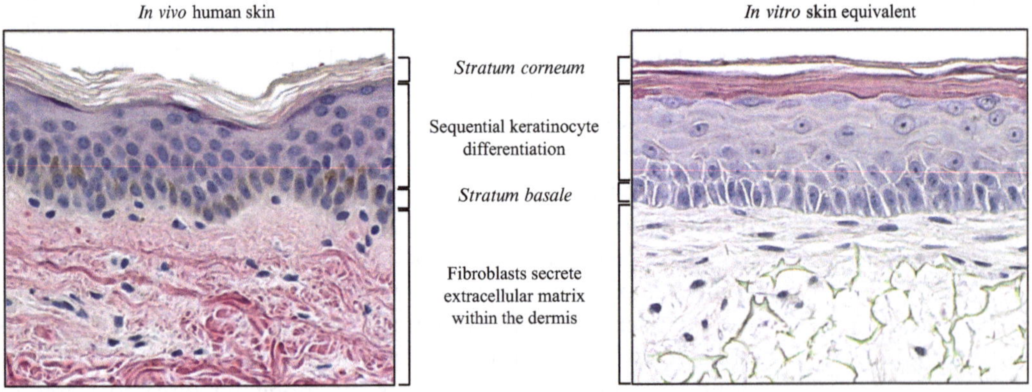

Fig. 3 Histological analysis of the full-thickness skin equivalent closely resembles human skin. Representative H&E micrographs of the skin equivalent, in comparison to healthy human skin. The skin equivalent resembles the morphology of human skin, and the self-assembled dermal compartment provides support for the successful differentiation and stratification of the epidermis. The skin sample is from the buttock site of a 23-year-old female, and provided by Procter and Gamble, Cincinnati, USA

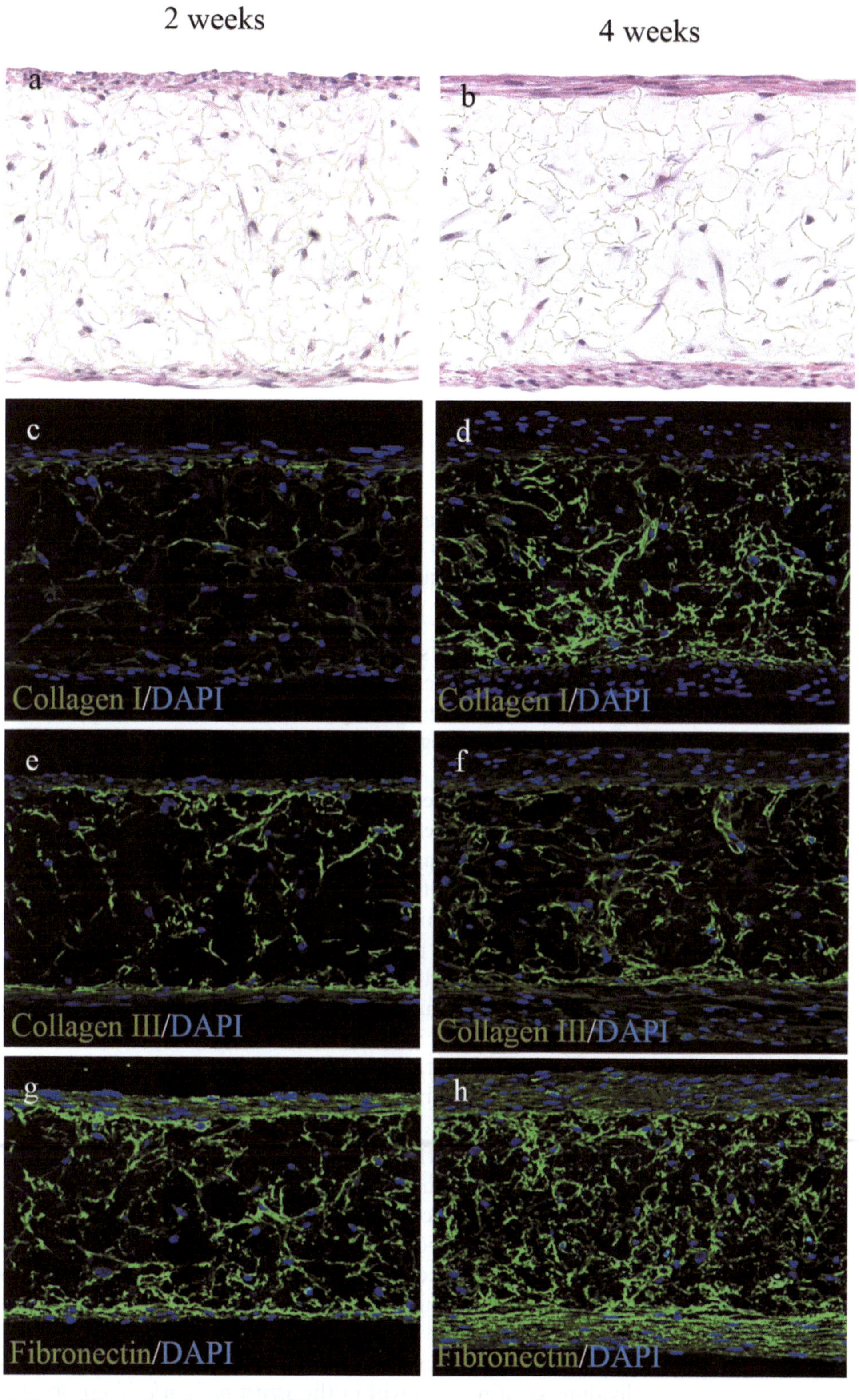

Fig. 4 Deposition of endogenous ECM over time. (**a**, **b**) Representative H&E micrographs of fibroblasts cultured within the Alvetex® scaffold for 2–4 weeks to generate a dermal equivalent. (**c**–**h**) Immunofluorescence analysis of collagen I, collagen III, and fibronectin deposited within the scaffold demonstrates increased ECM protein production over time

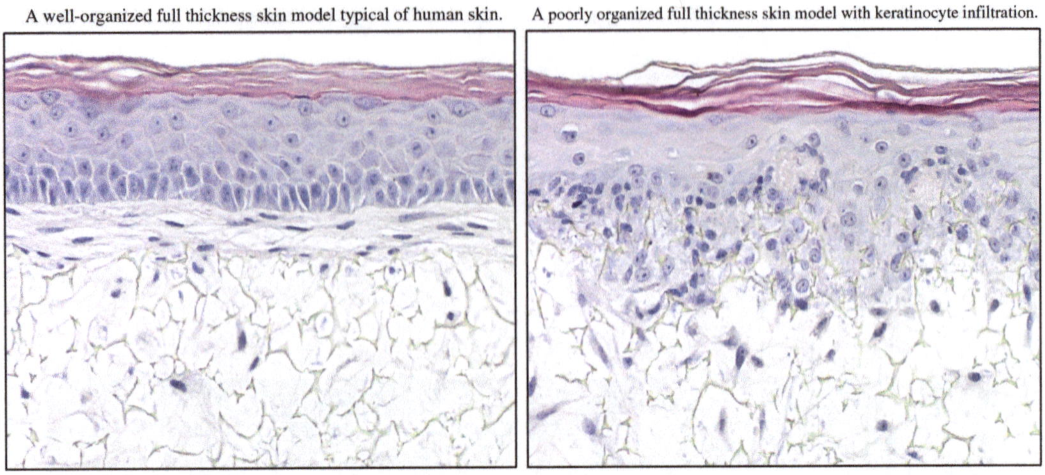

Fig. 5 Sufficient ECM protein production is essential to prevent keratinocyte infiltration. Representative H&E micrographs of full-thickness skin equivalents generated with mature and less developed dermal compartments in the absence and presence of keratinocyte infiltration, respectively

within the Alvetex® scaffold allows them to produce endogenous ECM, which can be visualized by analytical techniques such as immunofluorescence staining and electron microscopy [20]. The production of endogenous ECM proteins such as collagen and fibronectin increases over the culture period, and becomes deposited between the pores of the scaffold to form the dermal equivalent (Fig. 4). Due to the presence of endogenous ECM, the dermal equivalent serves as a robust assay for visualizing changes in ECM deposition. The length of the fibroblast culture period and amount of ECM protein deposition within the Alvetex® scaffold are important to fully support the epidermal compartment.

Histological and immunofluorescence analysis demonstrates that a mature dermal compartment with abundant ECM protein deposition is able to support a uniformly differentiated and stratified epidermis, in the absence of any keratinocyte infiltration (Fig. 5). If the fibroblasts have insufficient time to generate ECM proteins within the dermal equivalent, keratinocytes migrate and infiltrate into the scaffold, and they are unable to form an organized epidermis. The culture period of the dermal equivalent is therefore essential and must be optimized and characterized when incorporating different types of fibroblasts within the skin equivalent.

In the submerged phase of the protocol, the skin equivalent is cultured in low-calcium conditions to promote keratinocyte proliferation, which allows the formation of a layer of keratinocytes which express keratin 14, a prototypic marker of undifferentiated dividing keratinocytes within the *stratum basale*. To promote keratinocyte differentiation, the skin equivalents are raised to the

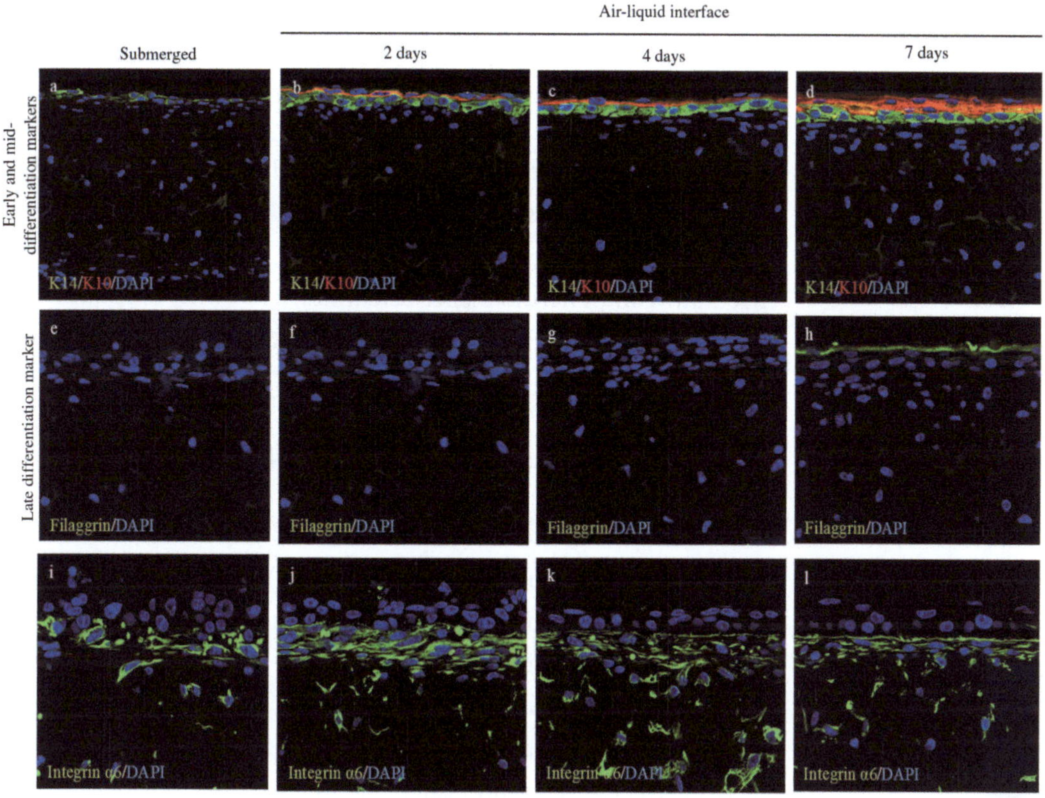

Fig. 6 Development of the epidermis over time. (**a-h**) Immunofluorescence staining of early, mid-, and late differentiation markers demonstrates the sequential differentiation and stratification to form the basal, suprabasal, and terminally differentiated keratinocytes within the first week of air-liquid interface culture. (**i-l**) Immunofluorescence staining of integrin α6 demonstrates basement membrane formation

air-liquid interface in high-calcium conditions, and keratinocytes differentiate into the *stratum spinosum* and *stratum granulosum* suprabasal layers within the first week, identified by the presence of keratin 10. After 7 days at the air-liquid interface, keratinocytes terminally differentiate into flattened corneocytes to form the *stratum corneum*, which is characterized by filaggrin (Fig. 6). The formation of the basement membrane was also demonstrated using the classical marker, integrin α6. Although this antibody also cross-reacts with integrins on the fibroblasts, a uniform line corresponding to the basement membrane is identified at the dermoepidermal junction from 4 days at the air-liquid interface (Fig. 6). Due to the absence of exogenous ECM, this basement membrane forms de novo as the epidermal and dermal components come together.

2 Materials

Human neonatal dermal fibroblasts (HDFn) (Thermo Fisher Scientific).

Human neonatal keratinocytes (HEKn) (Thermo Fisher Scientific).

12-Well Alvetex® insert (Reprocell).

6-Well plate (Greiner Bio-One).

Medium 106 (Thermo Fisher Scientific).

Low-serum growth supplement (LSGS) (Thermo Fisher Scientific).

EpiLife® (Thermo Fisher Scientific).

Human keratinocyte growth supplement (HKGS) (Thermo Fisher Scientific).

Gentamicin (Thermo Fisher Scientific).

Amphotericin B (Thermo Fisher Scientific).

Keratinocyte growth factor (KGF) (Thermo Fisher Scientific).

Calcium chloride (Sigma Aldrich).

TGFβ1 recombinant human protein (Thermo Fisher Scientific).

Ascorbic acid (Sigma Aldrich).

Trypsin-EDTA (Thermo Fisher Scientific).

Trypsin neutralizer (Thermo Fisher Scientific).

Sterile Dulbecco's phosphate-buffered saline (DPBS) (Scientific Laboratory Supplies Ltd.).

Sterile forceps.

3 Methods

3.1 Bioengineering Human Skin In Vitro

The full-thickness skin equivalent is developed using a step-by-step protocol, which involves the generation of a robust dermal equivalent by fibroblasts producing endogenous extracellular matrix proteins within the Alvetex® scaffold, followed by the generation of a well-organized stratified epidermal compartment (Fig. 7). This protocol was generated for a 12 well skin model, however Alvetex® scaffold is available in different formats, which can also be used (*see* **Note 1**). Examples of a mature dermal equivalent and full-thickness skin equivalent are shown in Figs. 8 and 9, respectively. All steps are to be carried out in a sterile, category 2 laminar flow hood at room temperature. When disposing of waste materials, follow your respective guidelines and COSHH (Control of Substances Hazardous to Health) forms (*see* **Note 2**).

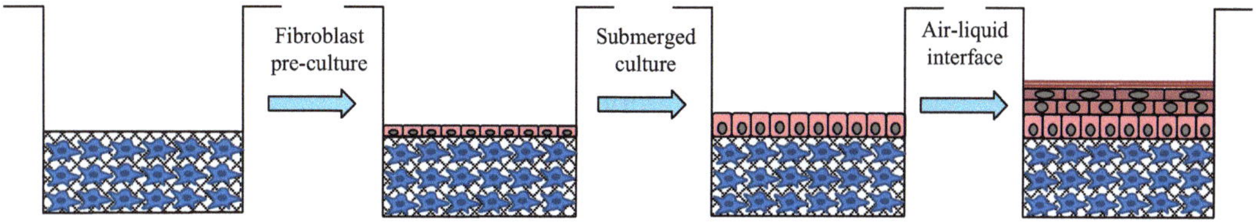

Fig. 7 Methodology used in the generation of the full-thickness skin equivalent. In the first phase, a dermal component is generated by seeding dermal neonatal fibroblasts into Alvetex® scaffold, and allowing them to establish and generate endogenous extracellular matrix proteins such as collagen and elastin. In the second phase, human neonatal keratinocytes are seeded onto the dermal component, and the full-thickness skin equivalent is sequentially cultured in submerged and air-liquid interface conditions

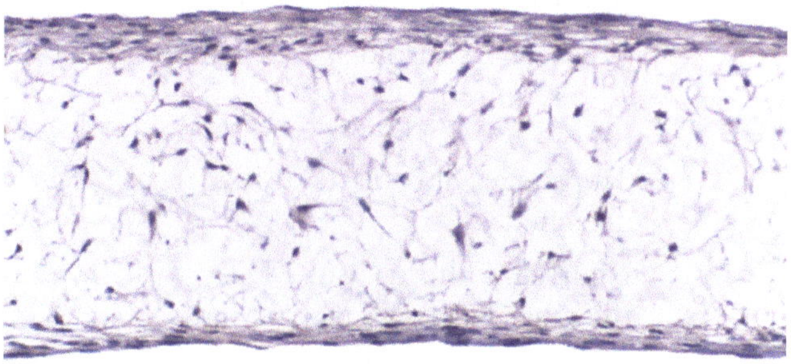

Fig. 8 Mature dermal equivalent. Representative H&E micrograph of a mature dermal equivalent demonstrating fibroblasts cultured in three-dimensions within the scaffold

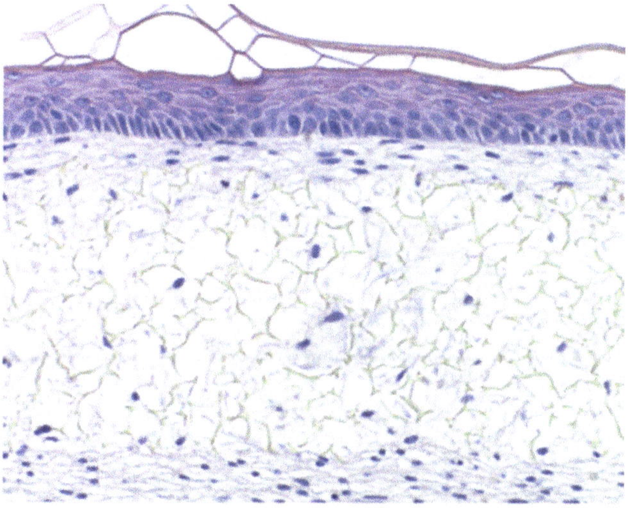

Fig. 9 Mature full-thickness skin equivalent. Representative H&E micrograph of a mature full-thickness skin equivalent, showing the multilayered structure of the epidermis supported by the underlying dermis

3.1.1 Revival of Fibroblasts

1. Prepare a bottle of complete Medium 106 (Medium 106 supplemented with 10 mL LSGS, 10 μg/mL gentamicin, and 0.25 μg/mL amphotericin B).
2. Revive and culture human dermal neonatal fibroblasts in complete Medium 106, as per the manufacturer's instructions.

3.1.2 Preparation of Alvetex® Scaffold

3. The Alvetex® scaffold must be treated with ethanol to render it hydrophilic before use (*see* **Note 3**). Pretreat 12-well Alvetex® scaffolds with:
   (a) 70% Ethanol.
   (b) 1× Sterile DPBS.
   (c) Complete Medium 106.
   (d) Leave scaffold submerged in complete Medium 106 until required.

3.1.3 Seeding of Fibroblasts onto the Alvetex® Scaffold

4. Trypsinize fibroblasts using trypsin-EDTA until cells have detached.
5. Neutralize the trypsin-EDTA with trypsin neutralizer, as per the manufacturer's instructions.
6. Centrifuge the cells at 200 × *g* for 5 min.
7. Perform cell counts using a trypan blue exclusion assay.
8. Seed 0.5 × 10^6 human dermal neonatal fibroblasts onto 12-well Alvetex® scaffold inserts.
9. Incubate at 37 °C in a 5% CO_2 humidified incubator in complete Medium 106 supplemented with 5 ng/mL TGF β1 and 100 μg/mL ascorbic acid.
10. Dermal equivalents can be maintained for up to 35 days.

3.1.4 Seeding of Keratinocytes onto the Dermal Equivalent

11. Prepare a bottle of complete EpiLife® media (EpiLife® supplemented with 5 mL HKGS, 10 μg/mL gentamicin, and 0.25 μg/mL amphotericin B).
12. Revive and culture human epidermal neonatal keratinocytes in complete EpiLife® media, as per the manufacturer's instructions.
13. Trypsinize keratinocytes using trypsin-EDTA until cells have detached.
14. Neutralize the trypsin-EDTA with trypsin neutralizer, as per the manufacturer's instructions.
15. Centrifuge the cells at 200 × *g* for 5 min.
16. Perform cell counts using a trypan blue exclusion assay.
17. Seed 1.3 ×10^6 human epidermal neonatal keratinocytes onto the mature dermal equivalent.

18. Incubate at 37 °C in a 5% CO_2 humidified incubator for 48 h in complete EpiLife® media supplemented with 10 ng/mL KGF, 140 μM $CaCl_2$, and 100 μg/mL ascorbic acid.

3.1.5 Raising Full-Thickness Models to the Air-Liquid Interface

19. After 48 h, raise the full-thickness skin equivalents to the air-liquid interface by carefully aspirating media from inside the insert.
20. Culture the full-thickness skin equivalents in complete EpiLife® media supplemented with 10 ng/mL KGF, 1.64 mM $CaCl_2$, and 100 μg/mL ascorbic acid.
21. After 7 days at the air-liquid interface, the epidermis is fully formed; however it can be maintained longer for long-term experiments.

3.2 Processing and Analysing the Skin Equivalents

HSEs must be characterised and validated prior to their use. The images obtained for this book chapter were generated using the following protocols for histological and immunofluorescence staining. Alternative protocols and reagents can also be used.

3.2.1 Processing Skin Equivalents for Paraffin Wax Embedding

1. Remove the skin equivalent from the plastic insert holders using a pair of flat-ended sterile forceps.
2. Wash the skin equivalent three times in 5 mL PBS, and fix the skin equivalents in 10% formalin for 2 h at room temperature, or overnight at 4 °C.
3. Wash the skin equivalents three times in 5 mL PBS.
4. Sequentially dehydrate the skin equivalents in 30%, 50%, 70%, 80%, 90%, and 95% ethanol v/v for 10 min each.
5. Place the skin equivalents in 100% ethanol for 30 min.
6. Transfer the skin equivalents to tissue-processing cassettes and place in Histoclear II for 30 min.
7. Incubate the skin equivalents in a 1:1 ratio of Histoclear II: molten paraffin wax in a convection oven at 65 °C for 30 min.
8. Place the skin equivalents in 100% molten wax for at least 1 h at 65 °C, prior to cutting the models in half across their diameter, and embedding in molten wax, with the flat edge facing down within the tissue-processing molds.
9. Allow the embedded samples to set at room temperature overnight prior to downstream applications.

3.2.2 Processing Skin Equivalents for Histological Analysis

1. Section the transverse skin equivalents at a thickness of 5 μm using a rotary manual microtome.
2. Place sections in a 37 °C water bath prior to mounting onto charged slides.
3. Allow slides to dry on a 30 °C heated slide rack before staining.
4. Deparaffinize the sections in Histoclear II for 5 min.

5. Sequentially rehydrate the skin equivalents in 95% ethanol, 70% ethanol, and distilled water for 1 min each.
6. Stain the slides with Mayer's hematoxylin for 5 min, which stains negatively charged components, such as nucleic acids, blue.
7. Wash away any excess Mayer's hematoxylin in distilled water for 30 s.
8. Transfer the samples to alkaline alcohol for 30 s to ensure that the hematoxylin dye appears blue.
9. Sequentially dehydrate the samples in 70% ethanol and 95% ethanol for 30 s each.
10. Stain with eosin for 30 s to stain positively charged components pink.
11. Further dehydrate the samples twice in 95% ethanol for 10 s, and then twice in 100% ethanol for 15–30 s, respectively.
12. Place the skin equivalents in Histoclear II twice for 3 min each.
13. Mount a clean coverslip onto the slide using Omnimount.
14. Leave to dry inside a fume hood at room temperature prior to imaging with a light microscope.

3.2.3 Processing Skin Equivalents for Immunofluorescence Analysis

1. Section the transverse skin equivalents at a thickness of 5 μm using a rotary manual microtome.
2. Place sections in a 37 °C water bath prior to mounting onto charged slides.
3. Allow slides to dry on a 30 °C heated slide rack prior to use.
4. Deparaffinize the sections in Histoclear II for 15 min.
5. Sequentially rehydrate the skin equivalents in 100% ethanol, 70% ethanol, and PBS for 5 min each.
6. Perform antigen retrieval in pH 6 citrate buffer at 95 °C for 20 min.
7. After cooling, block and permeabilize the slides in 150 μL 20% neonatal calf serum in 0.4% Triton X-100 in PBS for 1 h at room temperature.
8. Add 150 μL of the primary antibody diluted in blocking solution, and incubate with the samples overnight in a humidified chamber at 4 °C (Table 1).
9. Wash the samples three times in PBS for 10 min each.
10. Incubate the samples with 150 μL of the relevant secondary antibody diluted in blocking solution for 1 h at room temperature (Table 2).
11. Wash the samples three times in PBS for 10 min each.

Table 1
Primary antibodies used for immunofluorescence staining with suppliers and working concentrations

Antibody	Supplier	Product code	Dilution
Keratin 10	Abcam	ab76318	1:100
Keratin 14	Abcam	ab7800	1:100
Filaggrin	Abcam	ab17808	1:100
Collagen I	Abcam	ab34710	1:100
Collagen III	Abcam	ab7778	1:100
Fibronectin	Abcam	ab23750	1:100
Integrin 6	Abcam	ab181551	1:100

Table 2
Secondary antibodies used for immunofluorescence staining with suppliers and working concentrations

Antibody	Supplier	Product code	Dilution
Alexa-Fluor® 594 Donkey Anti-Rabbit IgG (H + L)	Thermo Fisher Scientific	A-21207	1:1000
Alexa-Fluor® 488 Goat Anti-Mouse IgG (H + L)	Thermo Fisher Scientific	A-11001	1:1000
Alexa-Fluor® 488 Donkey Anti-Rabbit IgG (H + L)	Thermo Fisher Scientific	A-21206	1:1000

12. Mount the coverslip onto the slides using Hardset Vectashield with DAPI.
13. Allow the Hardset Vectashield with DAPI to set for 15 min at room temperature, prior to imaging with a confocal microscope.

4 Notes

1. Alvetex® scaffold is commercially available in a range of formats, and additional insert sizes can be used to generate the full-thickness model, if cell seeding numbers are adjusted according to surface area.
2. All solutions and equipment coming into contact with the cells must be sterile and used within a Category 2 laminar flow hood. Aseptic technique should also be used accordingly.
3. Alvetex® scaffold is hydrophobic; therefore the initial ethanol wash is important to render it hydrophilic prior to the seeding of cells.

References

1. Proksch E, Brandner JM, Jensen JM (2008) The skin: an indispensable barrier. Exp Dermatol 17(12):1063–1072
2. Brenner M, Hearing VJ (2008) The protective role of melanin against UV damage in human skin. Photochem Photobiol 84(3):539–549
3. Merkel F (1875) Tastzellen und Tastkörperchen bei den Hausthieren und beim Menschen. Arch Mikrosk Anat 11(1): 636–652
4. Streilein JW, Bergstresser PR (1984) Langerhans cells: antigen presenting cells of the epidermis. Immunobiology 168(3–5):285–300
5. Eckert RL, Rorke EA (1989) Molecular biology of keratinocyte differentiation. Environ Health Perspect 80:109–116
6. Rice RH, Green H (1979) Presence in human epidermal cells of a soluble protein precursor of the cross-linked envelope: activation of the cross-linking by calcium ions. Cell 18(3):681–694
7. Briggaman RA, Wheeler CE (1975) The epidermal-dermal junction. J Investig Dermatol 65(1):71–84
8. Cotta-Pereira G, Rodrigo G, Bittencourt-Sampaio S (1976) Oxytalan, elaunin, and elastic fibers in the human skin. J Investig Dermatol 66(3):143–148
9. Wong VW, Sorkin M, Glotzbach JP, Longaker MT, Gurtner GC (2011) Surgical approaches to create murine models of human wound healing. Biomed Res Int 2011:969618
10. Gerber PA, Buhren BA, Schrumpf H, Homey B, Zlotnik A, Hevezi P (2014) The top skin-associated genes: a comparative analysis of human and mouse skin transcriptomes. Biol Chem 395(6):577–591
11. EU (2009) Regulation (EC) No. 1223/2009 of the European parliament and of the council of 30 November 2009 on cosmetic products (recast). Off J Eur Union L342:59–209
12. Bikle DD, Xie Z, Tu CL (2012) Calcium regulation of keratinocyte differentiation. Expert Rev Endocrinol Metab 7(4):461–472
13. Rehder J, Souto LRM, Issa CMBM, Puzzi MB (2004) Model of human epidermis reconstructed in vitro with keratinocytes and melanocytes on dead de-epidermized human dermis. Sao Paulo Med J 122(1):22–25
14. Carlson MW, Alt-Holland A, Egles C, Garlick JA (2008) Three-dimensional tissue models of normal and diseased skin. Curr Protoc Cell Biol:19–19
15. Mieremet A, Rietveld M, van Dijk R, Bouwstra JA, El Ghalbzouri A (2018) Recapitulation of native dermal tissue in a full-thickness human skin model using human collagens. Tissue Eng A 24(11–12):873–881
16. Knight E, Murray B, Carnachan R, Przyborski S (2011) Alvetex®: polystyrene scaffold technology for routine three dimensional cell culture. In: 3D cell culture. Humana Press, New York, NY, pp 323–340
17. Hill DS, Robinson ND, Caley MP, Chen M, O'Toole EA, Armstrong JL et al (2015) A novel fully humanized 3D skin equivalent to model early melanoma invasion. Mol Cancer Ther 14(11):2665–2673
18. Ng W, Ikeda S (2011) Standardized, defined serum-free culture of a human skin equivalent on fibroblast-populated collagen scaffold. Acta Derm Venereol 91(4):387–391
19. Reijnders CM, van Lier A, Roffel S, Kramer D, Scheper RJ, Gibbs S (2015) Development of a full-thickness human skin equivalent in vitro model derived from TERT-immortalized keratinocytes and fibroblasts. Tissue Eng A 21 (17–18):2448–2459
20. Roger M, Fullard N, Costello L, Bradbury S, Markiewicz E, O'Reilly S, Nelson G (2019) Bioengineering the microanatomy of human skin. J Anat 234(4):438–455

Chapter 10

Three-Dimensional Epidermal Model from Human Hair Follicle-Derived Keratinocytes

Takamitsu Matsuzawa, Michiyo Nakano, Ayako Oikawa, Yuumi Nakamura, and Hiroyuki Matsue

Abstract

Three-dimensional (3D) epidermal models reconstructed from human skin-derived keratinocytes have been utilized as an alternative to animal testing and models, not only in toxicology, but also in skin biology. Although there are currently several reconstructed human epidermis (RHE) models commercially available, the donors of the keratinocytes are not identified in these models. A tailor-made system is needed to investigate the individual differences in RHE derived from each donor.

It is possible to make an individual RHE using each donor's keratinocytes, which are usually obtained by invasive procedures such as skin excision or biopsy. To overcome this drawback, we established an RHE model using keratinocytes derived from plucked hair follicles as a less invasive procedure under conditions without feeder cells, serum, or matrix proteins. In this chapter, we provide a method of isolation and two-dimensional (2D) culture of keratinocytes derived from adult human plucked hair follicles including the outer root sheath (ORS). We also provide a detailed protocol for establishing an RHE model by culturing the keratinocytes under a 3D culture condition. We believe that our less invasive technique will provide a useful tool for investigating individual RHE in both normal and disease settings.

Key words Plucked hair follicle-derived keratinocytes, Three-dimensional reconstructed human epidermis, Individualization

1 Introduction

For decades, plucked hair follicles have been occasionally used as a source of keratinocytes in human studies. Plucked hair follicle-derived keratinocytes can be obtained in culture, either by direct outgrowth of keratinocytes derived from hair follicles [1–8] or by seeding cell suspensions obtained by enzymatic digestion of hair follicles [2, 5, 9–11]. In those protocols, feeder cells (i.e., fibroblasts), serum (i.e., fetal calf serum, fetal bovine serum, and autologous serum), and/or matrix proteins (e.g., basement membrane matrix or collagen) are generally required for primary cultures of plucked hair follicle-derived keratinocytes [1–3, 5–7,

Sophie Böttcher-Haberzeth and Thomas Biedermann (eds.), *Skin Tissue Engineering: Methods and Protocols*, Methods in Molecular Biology, vol. 1993, https://doi.org/10.1007/978-1-4939-9473-1_10,

9–11]. Therefore, we performed a primary culture of plucked hair follicle-derived keratinocytes under conditions without feeder cells, serum, or matrix proteins.

To date, many in vitro human three-dimensional (3D) skin models, known as skin equivalents, have been developed because of a recognition of the limitations of two-dimensional (2D) cultures for understanding skin biology, worldwide expansions of bans and limitations on animal experimentation, and challenge of treating chronic ulcers and large burn injuries [12, 13], and these models have been widely used as animal and human skin substitutes for skin research. As one type of skin models, reconstructed human epidermis (RHE) models closely mimic the biochemical and physiological properties of human epidermis. Currently, several commercially available RHE models have been used for various studies, such as metabolic studies of pharmaceutical products [14–16], determination of absorption properties [17–19], assessment of cutaneous corrosivity [20–22], and epidermal responses to irritants and sensitizers [23–26]. Although these RHE models are considered alternatives to animal testing, their major drawback is that they are derived from unknown donors, hampering studies of individual differences in RHE among donors. On the other hand, invasive procedures such as skin excision or biopsy are generally needed to obtain each donor's keratinocytes.

Previously, Limat et al. [2, 3] reported on the use of a reconstructed 3D epidermis using plucked hair follicle-derived keratinocytes. They used the epidermal equivalents as a less invasive treatment for chronic leg ulcers. Recently, the epidermal equivalents were used for evaluation of sunscreen genoprotection [7]. They cultured plucked hair follicle-derived keratinocytes under conditions with fetal calf serum, or autologous serum, in addition to feeder cells for 2D and 3D cultures. It is preferable to use simpler (without feeder cells) and safer (without serum) techniques when preparing epidermal equivalents. Therefore, our method establishes an RHE model by culturing plucked hair follicle-derived keratinocytes, which is a less invasive technique, under a 3D culture condition without feeder cells and serum. Our method will provide a useful tool for investigating individualized skin biology and studying epidermal barrier functions in patients with various skin disorders (e.g., atopic dermatitis or psoriasis).

2 Materials

2.1 Instruments and Supplies

1. Wide tweezers (*see* Fig. 1a).
2. Surgical tweezers (*see* Fig. 1b), which were sterilized via autoclave.
3. Surgical scissors (*see* Fig. 1c), which were sterilized via autoclave.

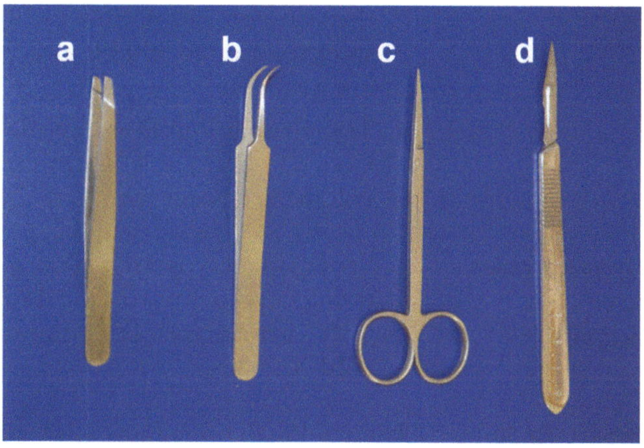

Fig. 1 The instruments used in this protocol. (**a**) Wide tweezers, (**b**) surgical tweezers, (**c**) surgical scissors, and (**d**) surgical knife No. 11

4. Non-tissue culture-treated plate, 24-well flat bottom.
5. Tissue culture plate, 24-well flat bottom.
6. Tissue culture plate, 6-well flat bottom.
7. Tissue culture dish, 100 × 20 mm.
8. Plastic petri dish, 100 × 15 mm.
9. ThinCert tissue culture insert for 24-well plates: Translucent polyethylene terephthalate (PET) membrane, 0.4 μm pore size, 33.6 mm^2 culture surface area. Place an insert into each well of a 24-well tissue culture plate in use.
10. Surgical knife No. 11 (*see* Fig. 1d).
11. Filter paper: When used, cut the paper to a size of approximately 3.5 cm × 4 cm.
12. Nylon mesh bag: 45 mm × 74 mm.

2.2 Isolation and 2D Culture of Hair Follicle-Derived Keratinocytes

1. CnT-PR medium (CELLnTEC Advanced Cell Systems, Bern, Switzerland): Store at −20 °C and thaw overnight at 4 °C before use. After thawing, store at 4 °C with protection from light for up to 6 weeks. In use at 37 °C, warm the necessary volume in a water bath. CnT-PR medium is completely free of animal- or human-derived components (http://cellntec.com/products/cnt-pr/#datasheet).
2. Kanamycin solution: 50 mg/mL Kanamycin from *Streptomyces kanamyceticus* in 0.9% NaCl, store at 4 °C, and dilute to a final concentration of 0.1 mg/mL for use in CnT-PR medium.
3. Dispase II solution: Dissolve and dilute Dispase II (Godo Shusei Co., Ltd., Tokyo, Japan) in CnT-PR medium to 10,000 PU/mL, divide to 1 mL volumes in sterile vials, and

then store at −20 °C until use. In use at 37 °C, dilute Dispase II stock solution to 500 PU/mL in PBS and warm the solution in a water bath.

4. Trypsin solution: Store TrypLE Select 1× (Gibco Invitrogen Co., NY, USA) at 4 °C. In use at 37 °C, warm the necessary volume in a water bath.

2.3 Cryopreservation of Cultured Keratinocytes

1. Serum-free cryopreservation media: CELLBANKER 2 (Nippon Zenyaku Kogyo Co., Ltd., Fukushima, Japan), store at 4 °C before use. For long-term storage, it was stored at −20 °C.

2.4 Medium for 3D Culture of Hair Follicle-Derived Keratinocytes

1. CnT-PR-3D medium (3D barrier medium, CELLnTEC): Store at −20 °C and thaw overnight at 4 °C before use. After thawing, store at 4 °C with protection from light for up to 6 weeks. In use at 37 °C, warm the necessary volume in a water bath. CnT-PR-3D medium is completely free of animal- or human-derived components (http://cellntec.com/products/cnt-pr-3d/#datasheet).

3 Methods

3.1 Isolation and Primary 2D Culture of Hair Follicle-Derived Keratinocytes

The method of isolation and 2D culture of primary keratinocytes from adult human plucked hair follicles was performed as reported previously, with minor modifications [4, 8].

1. Prepare a 100 × 15 mm plastic petri dish containing 5 mL of CnT-PR medium with antibiotic (4 °C).
2. Pluck a hair sample from the scalp of one adult donor using wide tweezers (*see* Fig. 1a) with one hand outside a clean bench. Transfer the hair follicle into 5 mL of CnT-PR medium with antibiotic (4 °C) in a petri dish while grasping the hair sample using wide tweezers (*see* **Note 1**). The hair sample should be in anagen growth phase with a visible ORS (*see* Fig. 2a and **Note 2**) and may also include the bulge region of the ORS.
3. Cut and remove the hair shaft 2–3 mm above the visible ORS using surgical scissors (*see* Fig. 1c) with another hand while grasping the hair sample using wide tweezers, and immerse the ORS in the medium.
4. Repeat Subheading 3.1, **steps 2–3**, for 6–10 hair samples per donor, and thereafter shake the petri dish containing the cut 6–10 hair samples several times.
5. Add 300 μL of Dispase II solution (500 PU/mL, 37 °C) to each well of a 24-well, non-tissue culture-treated plate in a clean bench.

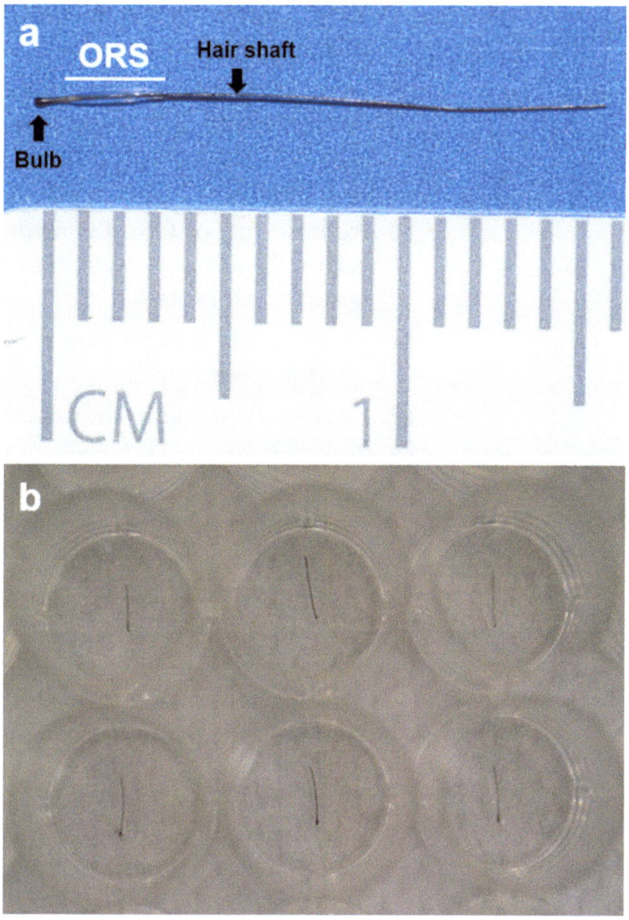

Fig. 2 A representative plucked hair sample from the occipital scalp of an adult donor and the attachment of hair follicles to the bottom of wells of a 24-well tissue culture plate. (**a**) The plucked hair sample includes a visible outer root sheath (ORS) (approximately 2–3 mm) along the hair shaft. (**b**) A hair follicle was placed in the center of each well without medium and attached to the bottom for 2 min

6. Transfer a hair follicle into each well of the 24-well plate containing Dispase II solution by grasping the cut short hair shaft above the visible ORS using surgical tweezers (*see* Fig. 1b and **Note 3**).
7. Incubate the 24-well plate at 37 °C in a humidified 5% CO_2 incubator for 5 min, aspirate the 300 μL of Dispase II solution, and add 300 μL of CnT-PR medium with antibiotic (4 °C) to each well of the 24-well plate (*see* **Note 4**).
8. Add 200 μL of CnT-PR medium with antibiotic (4 °C) to each well of a new 24-well tissue culture plate. Transfer a hair follicle into each well by grasping the hair shaft above the visible ORS

using surgical tweezers. Shake the new 24-well plate gently several times to wash the hair follicles.

9. Completely aspirate the medium, place a hair follicle in the center of each well using surgical tweezers, and then attach the hair follicles to the bottom of the wells for 2 min (*see* Fig. 2b and **Note 5**).
10. Add 165 μL of CnT-PR medium with antibiotic (37 °C) to each well of the 24-well plate using a sterile 200 μL micropipette tip (*see* **Note 6**).
11. Incubate the 24-well plate at 37 °C in a humidified 5% CO_2 incubator (*see* Fig. 3a).
12. Change the medium daily (*see* **Note** 7).
13. After the primary outgrowth (sparse keratinocytes) and subsequent proliferation from the ORS (*see* Fig. 3b, c), increase the volume of medium to 500 μL or 1 mL. Change the medium every 1–2 days (*see* **Note 8**).
14. The cells will proliferate and form a cobblestone appearance after 10–14 days of culture (*see* Fig. 3d). Even if the hair follicles are floating in the medium before formation of cobblestone appearance, the cells may proliferate and form a cobblestone appearance by changing the medium every 1–2 days.

3.2 Passage of Cultured Hair Follicle-Derived Keratinocytes

1. When the keratinocytes cover approximately 20–30% of a well surface of the 24-well tissue culture plate, aspirate the medium, wash the well with 300 μL of PBS (37 °C), and add 300 μL of TrypLE Select 1× (37 °C) to the well.
2. After incubation at 37 °C in a humidified 5% CO_2 incubator for 7 min, gently tap the 24-well plate while confirming cell detachment under a microscope.
3. After cell detachment, harvest the cells into a 15 mL centrifuge tube with 1–2 mL of CnT-PR medium with antibiotic (37 °C) to stop TrypLE activity by dilution.
4. Wash the cells by centrifugation at 1100 rpm (240 × *g*) for 5 min 2–3 times with CnT-PR medium with antibiotic (37 °C).
5. Resuspend the cell pellet in 1–1.5 mL of CnT-PR medium with antibiotic (37 °C), and then seed the cells into a well of a 6-well tissue culture plate.
6. Add additional 1–1.5 mL of CnT-PR medium with antibiotic (37 °C) to the 15 mL tube and harvest the medium into the well (2–3 mL per well).
7. Incubate the 6-well plate at 37 °C in a humidified 5% CO_2 incubator (**Passage 1**), 2 days later change the medium, and thereafter change the medium every 2–3 days (*see* Fig. 4a).

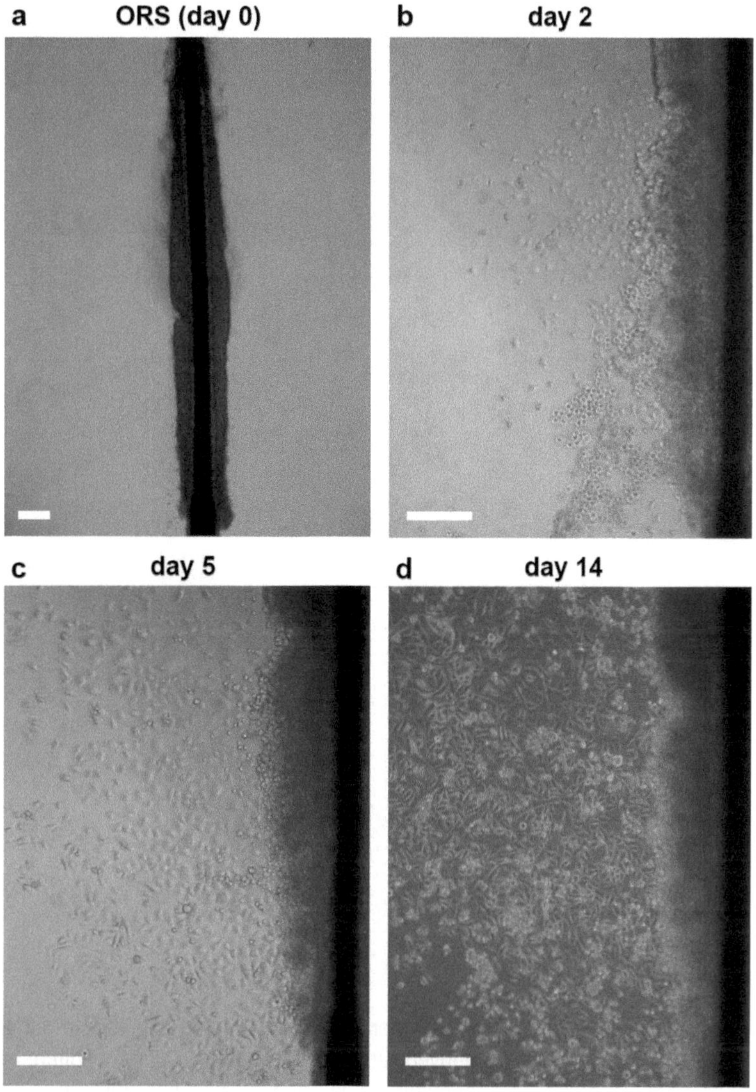

Fig. 3 Isolation of keratinocytes from plucked hair follicles including outer root sheath (ORS). (**a**) Representative microscopic image of hair follicles immediately after Dispase II treatment for 5 min at 37 °C (**day 0**). (**b**) Primary outgrowth of keratinocytes (sparse keratinocytes) was observed from ORS after 2–7 days of culture (**day 2**). (**c**) The keratinocytes proliferated (**day 5**) and (**d**) formed a cobblestone appearance after 10–14 days of culture (**day 14**). Scale bar, 200 μm

8. When the cells reach 60–90% confluence, aspirate the medium, wash the well with 1 mL of PBS (37 °C), and add 1 ml of TrypLE Select 1× (37 °C) to the well.
9. After incubation at 37 °C in a humidified 5% CO_2 incubator for 7 min, gently tap the 6-well plate while confirming cell detachment under a microscope.

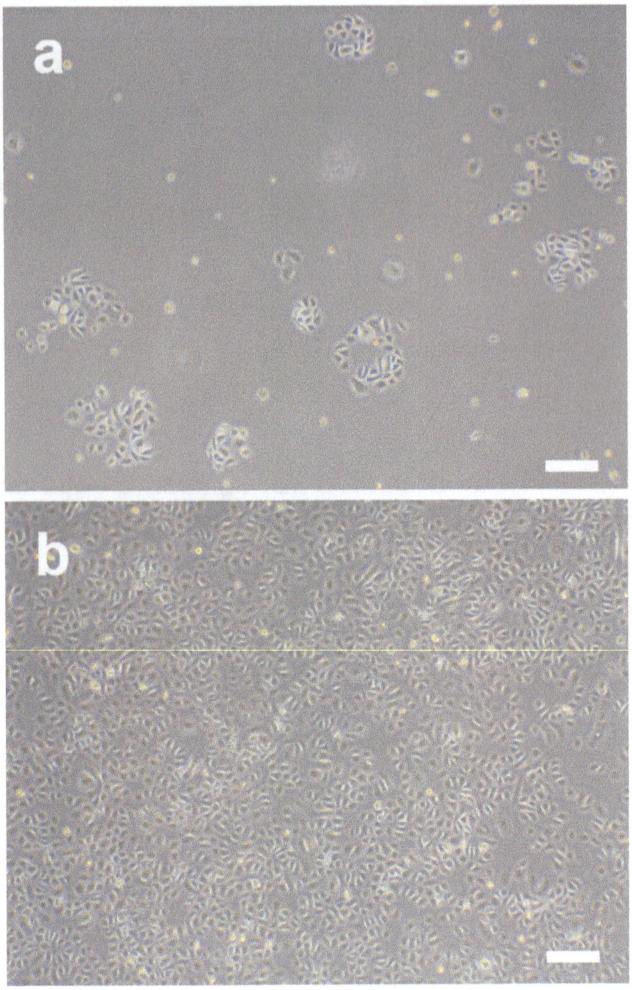

Fig. 4 Morphology of cultured hair follicle-derived keratinocytes. Hair follicle-derived keratinocytes (**a**) after 7 days of culture in passage 1 and (**b**) after 8 days of culture in passage 2. Scale bar, 200 μm

10. After cell detachment, harvest the cells into a 15 mL centrifuge tube with 4–5 mL of CnT-PR medium with antibiotic (37 °C).
11. Wash the cells by centrifugation at 1100 rpm for 5 min 2–3 times with CnT-PR medium with antibiotic (37 °C).
12. Resuspend the cell pellet in 3–5 mL of CnT-PR medium with antibiotic (37 °C), and then seed the cells into a 100 × 20 mm tissue culture dish.
13. Add additional 3–5 mL of CnT-PR medium with antibiotic (37 °C) to the 15 mL tube and harvest the medium into the tissue culture dish (6–10 mL per dish).

14. Incubate the tissue culture dish at 37 °C in a humidified 5% CO_2 incubator (**Passage 2**), 2 days later change the medium, and thereafter change the medium every 2–3 days (*see* Fig. 4b).
15. When the cells reach 60–90% confluence, the cells were harvested by incubation with 2 mL of TrypLE Select 1×, washed, and then seeded into two new 100 × 20 mm tissue culture dishes (**Passage 3**) as described in Subheading 3.2, **steps 8–14**.

At 60–90% confluence at **Passages 2–3**, the cells were either used to generate RHE (*see* Subheading 3.3) or suspended in CELL-BANKER 2 at 5×10^5 –1×10^6 cells/mL for frozen stocks stored at −80 °C until use.

3.3 Reconstruction of 3D Epidermis

The method of reconstruction of 3D epidermis described in the CELLnTEC 3D keratinocyte starter kit protocol (http://cellntec.com/products/resources/protocols/culture) was used, with minor modifications.

1. Place a culture insert (translucent PET membrane, 0.4 μm pore size, 33.6 mm^2 culture surface area) into each well of a 24-well tissue culture plate in a clean bench (*see* Fig. 5a, b).
2. Harvest the keratinocytes from a tissue culture dish by incubation with 2 mL of TrypLE Select 1× at 37 °C for 7 min. In these procedures, use CnT-PR medium without antibiotic (37 °C).
3. Wash the cells by centrifugation at 1100 rpm for 5 min 2–3 times with CnT-PR medium (37 °C).
4. Resuspend the cell pellet in CnT-PR medium (37 °C) and count the cells by trypan blue exclusion.
5. Separate the cells required for 3D culture, and increase the volume of medium to the necessary volume. For the 3D culture, 450 μL of cell suspension (1×10^5 cells) was added into an insert.
6. Add 450 μL of CnT-PR medium (37 °C) outside each insert, add 450 μL of cell suspension (1×10^5 cells) inside each insert, and then add additional 150 μL of CnT-PR medium (37 °C) outside each insert (total 600 μL outside) (*see* Fig. 5c).
7. Incubate the 24-well plate at 37 °C in a humidified 5% CO_2 incubator.
8. After incubation for 24 h, change the medium inside and outside each insert with 450–600 μL of CnT-PR medium (37 °C), respectively. Thereafter, change the medium daily (*see* **Note 9**).

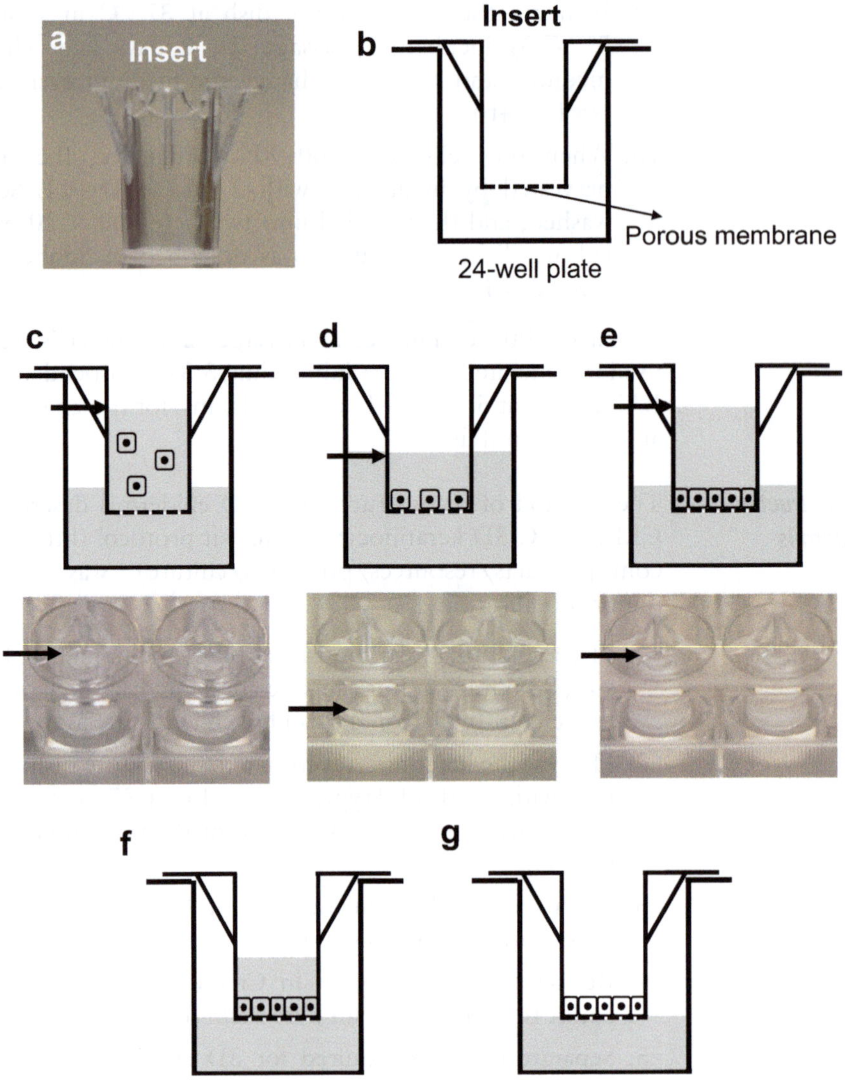

Fig. 5 The method of establishing a reconstructed human epidermis (RHE). (**a**) A photograph of a culture insert with a porous membrane (0.4 μm pore size). (**b**) A schematic drawing of a culture insert placed into a well of a 24-well plate. (**c**) 450 μL of CnT-PR medium was added outside each insert, 450 μL of cell suspension (1×10^5 cells) was added inside each insert, and then additional 150 μL of CnT-PR medium was added outside each insert (total 600 μL outside). (**d**) Until the cells reached confluence, the liquid level inside each insert was dropped to the level outside. (**e**) After the cells reached confluence, the liquid level inside each insert was higher than the level outside and maintained. (**c–e**) The arrow indicates the liquid level inside each insert. (**f**) For 3D culture, 200 μL of CnT-PR-3D medium was added inside each insert and 450 μL of CnT-PR-3D medium was added outside each insert. (**g**) 12 h later, the medium inside and outside each insert was aspirated and the cells were exposed to the air–liquid interface by adding 450 μL of CnT-PR-3D medium only outside each insert

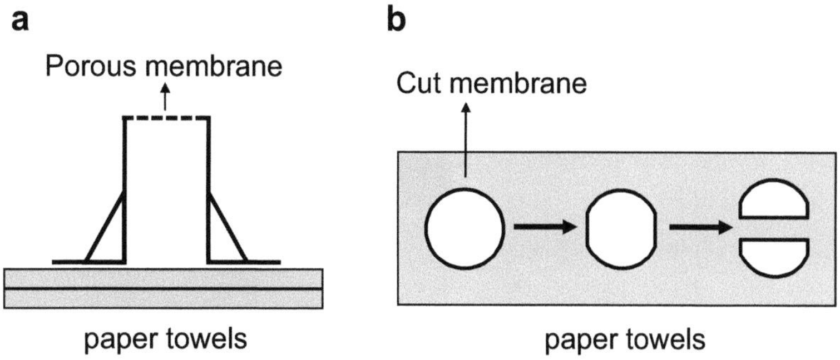

Fig. 6 The method of fixation and cutting of RHE. After fixation of RHE with 10% formalin in each well of a 24-well plate at 4 °C overnight, (**a**) an insert was placed on paper towels with the lower surface of a porous membrane facing upward. (**b**) The membrane, cut with a surgical knife No. 11, was placed on paper towels with the upper surface of the membrane facing upward and then divided with a surgical knife No. 11. In subsequent processes, *see* **Note 11**

9. After 100% confluence, transfer an insert into each well of a new 24-well tissue culture plate.
10. Aspirate the medium inside each insert, add 200 μL of CnT-PR-3D medium (37 °C) inside each insert, and then add 450 μL of CnT-PR-3D medium (37 °C) outside each insert (*see* Fig. 5f).
11. Incubate the 24-well plate at 37 °C in a humidified 5% CO_2 incubator.
12. After incubation for 12 h, aspirate the medium inside and outside each insert.
13. Expose the cells to the air–liquid interface by adding 450 μL of CnT-PR-3D medium (37 °C) only outside each insert (*see* Fig. 5g).
14. Incubate the 24-well plate at 37 °C in a humidified 5% CO_2 incubator (3D culture).
15. Change the medium outside each insert with 450 μL of CnT-PR-3D medium (37 °C) daily (*see* **Note 10**).
16. 14 Days after 3D culture, the RHE was fixed in 10% formalin at 4 °C for at least 24 h (*see* Fig. 6a, b and **Note 11**), embedded in paraffin, and observed by H&E staining (*see* Fig. 7a, b). 14 Days after 3D culture, reconstruction of 3D epidermis was achieved (*see* **Note 12**).

4 Notes

1. Hair follicles should be immediately transferred into the medium to avoid the drying out of the cells on them.

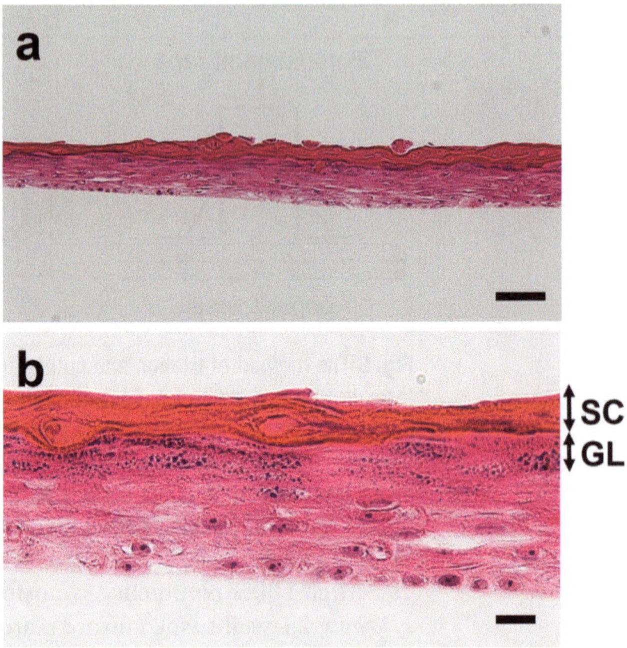

Fig. 7 A representative H&E section of RHE. Hair follicle-derived keratinocytes were cultured on a porous membrane at the air–liquid interface (3D culture). A representative H&E section of RHE is shown 14 days after 3D culture. (**a**) H&E staining, 10×. Scale bar, 50 μm. (**b**) H&E staining, 40×. Scale bar, 10 μm. The RHE was comprised of the stratified epidermal layers, including the granular layer (GL) and stratum corneum (SC). The RHE was often detached from the membrane in the process of fixation and embedding

2. The occipital part of the head is particularly suitable, giving many hairs in anagen growth phase with a large amount of ORS cells on the hair shaft [5].
3. At the time of transfer of the collected hair follicles, grasping the cut short hair shafts above the visible ORS by surgical tweezers may help avoid damage to the ORS.
4. If outgrowth of keratinocytes from the hair follicles is not observed within 7 days, it may be preferable to retry incubation at 37 °C for 15 min instead of 5 min for the same donor.
5. The attachment of hair follicles to the bottom of wells is crucial for outgrowth of keratinocytes from them, and may help avoid floating hairs at the time of medium addition.
6. Slowly add one drop of the medium to the hair follicle, and then add the remaining medium slowly from the edge of the bottom of the well. Take care not to float the hair follicles in the medium.
7. Carefully aspirate the medium and gently add 165 μL of fresh CnT-PR medium with antibiotic (37 °C) from the edge of the bottom of each well.

8. After 2–7 days of culture, primary outgrowth (sparse keratinocytes) was observed from the ORS. We usually observe outgrowth in 1–2 (out of approximately 6) hair follicles. In some hair follicles, we also observed outgrowth of keratinocytes from the bulge region of the ORS. Gently change the medium from the side of each well. It may be preferable for keratinocyte proliferation that the hair follicles are attached to the bottom of wells.
9. When replacing the medium, aspirate the CnT-PR medium outside each insert and then aspirate CnT-PR medium inside each insert. Subsequently, add 450 μL of CnT-PR medium (37 °C) inside each insert and then add 600 μL of CnT-PR medium (37 °C) outside each insert. The liquid level inside each insert is dropped to the level outside until the cells reach 100% confluence (*see* Fig. 5d). Especially note that the liquid level inside each insert is higher than the level outside, meaning 100% confluence (*see* Fig. 5e). The culture period is typically 2–3 days. If 100% confluence is not achieved after 3 days, the cell growth is unlikely to be sufficient to successfully develop a 3D culture.
10. Take care not to trap air bubbles underneath the porous membrane for the duration of the 3D culture. The inside surface of each insert should be dry for the 3D culture. If the outside medium seeps into the edge of the porous membrane, aspirate the seeping medium (approximately 10–20 μL) at the time of medium change.
11. 14 Days after 3D culture, an insert was transferred into each well of a new 24-well plate. 10% Formalin (4 °C) was added inside and outside each insert, and the 24-well plate was incubated at 4 °C overnight. Thereafter, an insert was placed on paper towels with the lower surface of the porous membrane facing upward (*see* Fig. 6a), and the membrane was cut with a surgical knife (No. 11) along the edge of the membrane. The cut membrane was placed on paper towels with the upper surface of the membrane facing upward, and then divided with the surgical knife (*see* Fig. 6b). Each of the divided membranes was placed on a filter paper (approximately 3.5 cm × 4 cm) with the upper surface of the membrane facing upward. Each filter paper was put into a nylon mesh bag and then the nylon mesh bag was immersed in 10% formalin at 4 °C until embedding in paraffin of RHE.
12. The RHE established with the CnT-PR-3D medium showed more keratinization (*see* Fig. 7a, b) than that established with the previous CnT-02-3DP medium [8], according to the commercial information of the CnT-PR-3D medium (http://cellntec.com/products/cnt-pr-3d/).

References

1. Imcke E, Mayer-da-Silva A, Detmar M, Tiel H, Stadler R, Orfanos CE (1987) Growth of human hair follicle keratinocytes in vitro. Ultrastructural features of a new model. J Am Acad Dermatol 17:779–786
2. Limat A, Mauri D, Hunziker T (1996) Successful treatment of chronic leg ulcers with epidermal equivalents generated from cultured autologous outer root sheath cells. J Invest Dermatol 107:128–135
3. Limat A, French LE, Blal L, Saurat JH, Hunziker T, Salomon D (2003) Organotypic cultures of autologous hair follicle keratinocytes for the treatment of recurrent leg ulcers. J Am Acad Dermatol 48:207–214
4. Sasahara Y, Yoshikawa Y, Morinaga T, Nakano Y, Kanazawa N, Kotani J, Kawamata S, Murakami Y, Takeuchi K, Inoue C, Kitano Y, Hashimoto-Tamaoki T (2009) Human keratinocytes derived from the bulge region of hair follicles are refractory to differentiation. Int J Oncol 34:1191–1199
5. Aasen T, Izpisúa Belmonte JC (2010) Isolation and cultivation of human keratinocytes from skin or plucked hair for the generation of induced pluripotent stem cells. Nat Protoc 5:371–382
6. Guiraud B, Hernandez-Pigeon H, Ceruti I, Mas S, Palvadeau Y, Saint-Martory C, Castex-Rizzi N, Duplan H, Bessou-Touya S (2014) Characterization of a human epidermis model reconstructed from hair follicle keratinocytes and comparison with two commercially models and native skin. Int J Cosmet Sci 36:485–493
7. Bacqueville D, Douki T, Duprat L, Rebelo-Moreira S, Guiraud B, Dromigny H, Perier V, Bessou-Touya S, Duplan H (2015) A new hair follicle-derived human epidermal model for the evaluation of sunscreen genoprotection. J Photochem Photobiol B 151:31–38
8. Nakano M, Kamada N, Suehiro K, Oikawa A, Shibata C, Nakamura Y, Matsue H, Sasahara Y, Hosokawa H, Nakayama T, Nonaka K, Ohara O (2016) Establishment of a new three-dimensional human epidermal model reconstructed from plucked hair follicle-derived keratinocytes. Exp Dermatol 25:903–906
9. Limat A, Noser FK (1986) Serial cultivation of single keratinocytes from the outer root sheath of human scalp hair follicles. J Invest Dermatol 87:485–488
10. Limat A, Hunziker T, Boillat C, Bayreuther K, Noser F (1989) Post-mitotic human dermal fibroblasts efficiently support the growth of human follicular keratinocytes. J Invest Dermatol 92:758–762
11. Moll I (1995) Proliferative potential of different keratinocytes of plucked human hair follicles. J Invest Dermatol 105:14–21
12. Auxenfans C, Fradette J, Lequeux C, Germain L, Kinikoglu B, Bechetoille N, Braye F, Auger FA, Damour O (2009) Evolution of three dimensional skin equivalent models reconstructed in vitro by tissue engineering. Eur J Dermatol 19:107–113
13. Ali N, Hosseini M, Vainio S, Taïeb A, Cario-André M, Rezvani HR (2015) Skin equivalents: skin from reconstructions as models to study skin development and diseases. Br J Dermatol 173:391–403
14. Gysler A, Kleuser B, Sippl W, Lange K, Korting HC, Höltje HD, Korting HC (1999) Skin penetration and metabolism of topical glucocorticoids in reconstructed epidermis and in excised human skin. Pharm Res 16:1386–1391
15. Stinchcomb AL (2003) Xenobiotic bioconversion in human epidermis models. Pharm Res 20:1113–1118
16. Hu T, Bailey RE, Morrall SW, Aardema MJ, Stanley LA, Skare JA (2009) Dermal penetration and metabolism of p-aminophenol and p-phenylenediamine: application of the EpiDerm human reconstructed epidermis model. Toxicol Lett 188:119–129
17. Gabbanini S, Lucchi E, Carli M, Berlini E, Minghetti A, Valgimigli L (2009) In vitro evaluation of the permeation through reconstructed human epidermis of essentials oils from cosmetic formulations. J Pharm Biomed Anal 50:370–376
18. Kurasawa M, Kuroda S, Kida N, Murata M, Oba A, Yamamoto T, Sasaki H (2009) Regulation of tight junction permeability by sodium caprate in human keratinocytes and reconstructed epidermis. Biochem Biophys Res Commun 381:171–175
19. Frasch HF, Dotson GS, Barbero AM (2011) In vitro human epidermal penetration of 1-bromopropane. J Toxicol Environ Health A 74:1249–1260
20. Hoffmann J, Heisler E, Karpinski S, Losse J, Thomas D, Siefken W, Ahr HJ, Vohr HW, Fuchs HW (2005) Epidermal-skin-test 1,000 (EST-1,000)--a new reconstructed epidermis for in vitro skin corrosivity testing. Toxicol In Vitro 19:925–929
21. Kandárová H, Liebsch M, Spielmann H, Genschow E, Schmidt E, Traue D, Guest R,

Whittingham A, Warren N, Gamer AO, Remmele M, Kaufmann T, Wittmer E, De Wever B, Rosdy M (2006) Assessment of the human epidermis model SkinEthic RHE for in vitro skin corrosion testing of chemicals according to new OECD TG 431. Toxicol In Vitro 20:547–559

22. Tornier C, Roquet M, Fraissinette Ade B (2010) Adaptation of the validated SkinEthic Reconstructed Human Epidermis (RHE) skin corrosion test method to 0.5 cm2 tissue sample. Toxicol In Vitro 24:1379–1385

23. Portes P, Grandidier MH, Cohen C, Roguet R (2002) Refinement of the Episkin protocol for the assessment of acute skin irritation of chemicals: follow-up to the ECVAM prevalidation study. Toxicol In Vitro 16:765–770

24. Coquette A, Berna N, Vandenbosch A, Rosdy M, De Wever B, Poumay Y (2003) Analysis of interleukin-1alpha (IL-1alpha) and interleukin-8 (IL-8) expression and release in in vitro reconstructed human epidermis for the prediction of in vivo skin irritation and/or sensitization. Toxicol In Vitro 17:311–321

25. Borlon C, Godard P, Eskes C, Hartung T, Zuang V, Toussaint O (2007) The usefulness of toxicogenomics for predicting acute skin irritation on in vitro reconstructed human epidermis. Toxicology 241:157–166

26. Lu B, Miao Y, Vigneron P, Chagnault V, Grand E, Wadouachi A, Postel D, Pezron I, Egles C, Vayssade M (2017) Measurement of cytotoxicity and irritancy potential of sugar-based surfactants on skin-related 3D models. Toxicol In Vitro 40:305–312

Chapter 11

Fabrication of a Co-Culture System with Human Sweat Gland-Derived Cells and Peripheral Nerve Cells

Matthias Brandenburger and Charli Kruse

Abstract

The interaction of peripheral nerves with different cells of the skin is a relevant aspect of many physiological processes including nociception, temperature control, and wound healing. Here we describe a protocol for the setup of an indirect co-culture system of peripheral nerve cells and sweat gland-derived stem cells, which can be used to quantify neurite outgrowth.

Key words Peripheral nerves, Stem cells, Indirect co-culture, Wound healing, Nerve regeneration, Sweat glands

1 Introduction

In the past, isolated peripheral nerve cells from rat dorsal root ganglia (DRG) were extensively used to analyze physiology [1–3] and regeneration [4, 5] of peripheral nerves. Recently, innervated skin models have successfully been implemented to address current questions of skin physiology [6–8].

The interaction between skin stem cells and peripheral nerves is a relevant topic, since sweat glands [9, 10] and peripheral nerves [11] play important roles in wound regeneration. For further analysis of interactions between peripheral nerve cells and sweat gland-derived stem cells, we developed an indirect co-culture model [12]. This model system can be used for the analysis of overall neurite outgrowth in indirect co-culture experiments. The co-culture system is set up by using cell-impermeable culture inserts in 24-well plates. This results in two separated cultivation chambers, which allow for the indirect co-cultivation of sweat gland-derived stem cells and peripheral nerve cells. After 24 h, the co-cultivation is stopped. Neurite outgrowth can be analyzed by

Sophie Böttcher-Haberzeth and Thomas Biedermann (eds.), *Skin Tissue Engineering: Methods and Protocols*, Methods in Molecular Biology, vol. 1993, https://doi.org/10.1007/978-1-4939-9473-1_11,

staining for neuron-specific markers and examination of neuronal structures under the microscope. Quantification of neurite length can subsequently be performed by using a freely available tool (VesSeg Tool), which has originally been developed for vessel segmentation.

We used this model system to demonstrate positive effects of glandular stem cells on peripheral nerve cells [12]. However, the following protocol may easily be adapted to address different questions in human skin physiology.

2 Materials

2.1 Isolation Media

1. Isolation of dorsal root ganglion (DRG) neurons:
   (a) Collagenase and dispase solution: 5 mL Sterile MEM Earle's, 1 mg/mL 0.92 U/mg collagenase, 2 mg/mL 0.99 U/mg Dispase II.
   (b) Trypsin solution: 0.1% (v/v) Trypsin in PBS.
2. Isolation of hSGSCs (sweat gland-derived stem cells):
   (a) HEPES solution: 2383 g HEPES in 100 mL Aqua bidest., adjust pH to 7.6.
   (b) Dispase-containing isolation medium: 13.33 mg Dispase II (2 U/mL) in 6 mL HEPES-solution,
   (c) HEPES-Eagle-medium: 90 mL Modified Eagle medium, 10 mL HEPES solution, adjust pH to 7.4.
   (d) Collagenase-containing isolation medium: 32 mL HEPES-Eagle-medium, 8 mL of 5% BSA in Aqua bidest., 300 µL 0.1 M calcium chloride, 8 mL 0.92 U/mg collagenase, adjust pH to 7.4.

2.2 Cultivation Media

1. DRG neurons:
   (a) Neurobasal A medium (Invitrogen): 5 mL Neurobasal A medium, 2 mM L-glutamine, 100 U/mL penicillin, 0.1 mg/mL streptomycin, 100 µL B-27 (Invitrogen), 50 ng/mL NGF 2.5 S.
2. hSGSCs:
   (a) Cultivation medium: 100 U/mL Penicillin, 0.1 mg/mL streptomycin in standard DMEM; depending on the kind of cultivation medium that is needed add 10% or 20% of FCS (*see* **Note 1**).

2.3 Solutions for Immunostaining

1. TBST stock solution: 8.76 g 150 mM Sodium chloride, 1.21 g 10 mM Tris, 500 µL Triton X-100 (0.05%), 1 g BSA (0.1%) in 1 L distilled water, adjust pH to 8.0.

2.4 Primary Antibodies

1. Dissolve neurofilament antibodies in TBST stock solution to the desired concentration (e.g., 2:1 mix of monoclonal NF-L (neurofilament light) and NF-H&M (neurofilament heavy and medium) antibodies, 1:200).

2.5 Secondary Antibody

1. Dissolve appropriate secondary antibody in PBS to the desired concentration (e.g., FITC-labeled anti-mouse-IgG 1:500).

3 Methods

3.1 Cell Preparation

1. In contrast to peripheral nerve cells, sweat gland-derived stem cells (hSGSCs) can easily be propagated and cryopreserved. Therefore, it is advisable to isolate a cryopreserved stock of hSGSCs before the preparation of the co-culture experiment begins.

3.2 Isolation and Propagation of hSGSCs

1. Cut the skin into small pieces using a scalpel.
2. Digest the skin overnight at 4 °C with dispase-containing isolation medium.
3. Remove the epidermis and incubate the dermis in collagenase-containing isolation medium for 3 h at 37 °C under constant shaking.
4. Filter the tissue through a nylon mesh and transfer the tissue into cultivation medium.
5. Stain the tissue with neutral red and sort out sweat glands by hand under the microscope.
6. Transfer the sweat glands into a type IV collagen-coated cell culture dish.
7. Immobilize sweat glands by scratching them into the bottom of the culture dish by cannulae.
8. Cultivate sweat glands using cultivation media with 20% FCS.
9. After outgrowing hSGSCs reach confluence, subculture cells using 0.1% trypsin and reduce FCS concentration to 10% for further cultivation.
10. Change medium every 3–4 days.
11. Cryopreserve the cells at the desired passage (*see* **Note 2**).

3.3 Poly-L-Lysine Coating of Coverslips

1. Wash coverslips (13 mm diameter) in 70% ethanol for at least 1 min.
2. Wet a cloth with ethanol and place it under a sterile hood.
3. Place the washed coverslip onto the wet cloth and let it dry.
4. Pipette 20 μL poly-L-lysine hydrobromide (PLL, mol wt ≥300,000, 0.1 mg/mL) into sterile petri dish.

5. Place the prepared coverslips onto the drops using forceps.
6. Carefully press coverslips down.
7. Incubate for 30 min at room temperature.
8. Loosen coverslips by carefully pipetting sterile water at the edge of the coverslips using a 1000 μL pipette.
9. Carefully remove coverslips by lifting the edge using forceps.
10. Wash the coverslips with sterile water.
11. Place coverslips onto sterile petri dish and let it dry for 2 h.
12. When sealed with parafilm, the coverslips can be stored in the petri dish at room temperature for up to 4 weeks (*see* **Note 3**).

3.4 Laminin Coating of Coverslips

1. Preheat laminin in water bath to 37 °C.
2. Dissolve laminin in PBS (1:20).
3. Pipette 20 μL laminin (50 μg/mL) into sterile petri dish.
4. Place the poly-L-lysine-coated coverslips onto the drops using forceps, the precoated side facing down.
5. Carefully press coverslips down.
6. Incubate for 35 min at 37 °C.
7. Loosen coverslips by carefully pipetting sterile water at the edge of the coverslips using a 1000 μL pipette.
8. Carefully remove coverslips by lifting the edge using forceps.
9. Place the coverslips in individual wells of a 24-well plate, the coated side facing up.
10. Carefully wash the coverslips in the 24-well plate with sterile MEM.
11. Cover each well with a sufficient amount of MEM and store the culture plate in the incubator until use (up to 24 h).

3.5 Preparation of Dorsal Root Ganglia (DRG)

In the past, isolated DRG neurons from rats were used as an in vitro model for peripheral nerves [1–3, 13]. However, human iPS-derived peripheral nerve cells recently became commercially available [14, 15]. In the future, stem cell-derived human peripheral nerves will assumably replace rodent DRG preparations. The following protocol gives a brief overview on the isolation of DRG neurons from adult or prenatal rodents. However, steps may need to be adapted for local ethic committee approval.

3.5.1 Adult Rats

1. Sacrifice rat (e.g., Wistar rat aged 50–70 days) by CO_2 asphyxiation immediately before preparation.
2. Open vertebral column from the thoracic to lumbar region.
3. Collect dorsal root ganglia (DRG) in Leibovitz-15 (L-15) medium at 4 °C.

4. Remove connective tissue and wash the DRG twice in PBS.
5. Store DRG in Leibovitz-15 (L-15) medium at 4 °C until enzymatic digestion.
6. Transfer DRG into isolation medium containing collagenase and dispase.
7. Mechanically mince DRG with scissors for 1 min.
8. Incubate tissue for 10 min at 37 °C under constant shaking (150 cycles/min).
9. Repeat the two previous steps for a second time.
10. Resuspend the cell solution using a 1000 μL pipette 50 times.
11. Filter the solution through a 100 μm cell strainer and centrifuge at 180 × *g* for 5 min (*see* **Note 4**).
12. Dissolve the pellet with 1500 μL MEM.

3.5.2 Prenatal Rats

1. Sacrifice dam by CO_2 asphyxiation immediately before preparation.
2. Remove prenatal rats from uterus and collect them in L-15 medium at 4 °C.
3. Prepare vertebral columns and part dorsal and ventral side.
4. Collect dorsal root ganglia (DRG) in Leibovitz-15 (L-15) medium at 4 °C.
5. Remove connective tissue and wash the DRG twice in PBS.
6. Store DRG in Leibovitz-15 (L-15) medium at 4 °C until enzymatic digestion.
7. Transfer DRG into isolation medium containing collagenase and dispase or use trypsin isolation medium instead.
8. Mechanically mince DRG with scissors for 1 min.
9. Incubate tissue for 10 min at 37 °C under constant shaking (150 cycles/min).
10. Repeat the two previous steps for a second time.
11. Resuspend the cell solution using a 1000 μL pipette 50 times.
12. Filter the solution through a 100 μm cell strainer and centrifuge at 180 × *g* for 5 min (*see* **Note 4**).
13. Dissolve the pellet with 1500 μL MEM.

3.6 Setup of the Indirect Co-culture System

1. Prepare hSGSC cultures from the cryostock 1 day prior to DRG isolation.
2. Seed hSGSCs in cell culture insert (*see* **Note 5**) using a density of 2.0×10^4 cells/cm^2.
3. Use hSGSC cultivation medium with 10% FCS and cultivate the hSGSCs for 1 day.

4. Seed DRG neurons onto pre-coated glass coverslips in 24-well plates using a density of 4–5 × 10^4 neurons/cm^2 (*see* **Note 6**).
5. Use 1:1 mixtures of hSGSC medium without FCS and neuronal media (*see* **Note 7**).
6. In this mixture use neurobasal A medium for adult DRG neurons and neurobasal medium for prenatal DRG neurons.
7. Start the indirect co-culture experiment by placing SGSC containing cell culture inserts into previously prepared neuron containing wells of 24-well plates (*see* Fig. 1).
8. Usually the indirect co-culture is stopped after 24 h, because this period still allows for the identification of individual neurons and its neurites. However, if you want to extend the co-cultivation period, change 50% of the culture media each day.

3.7 Data Acquisition and Analysis

1. Use a mixture of antibodies against different neurofilaments.
2. Take pictures on current fluorescent microscopes of immunostained specific neuronal structures for the clear identification of neurites.
3. The fluorescent pictures offer an appropriate contrast, which is needed for further computational quantification of neurite outgrowth.

3.8 Immunocyto-chemistry

1. Wash cells with PBS.
2. Fix cells with 4% paraformaldehyde and 10% sucrose in PBS for 10 min at 20 °C.
3. Wash three times with PBS.
4. Permeabilize with 0.1% Triton X-100 in PBS containing 1 mg/mL DAPI for 10 min at 20 °C.
5. Wash three times with PBS.
6. Incubate with 1.65% normal goat serum for 20 min at 20 °C to prevent nonspecific binding.
7. Incubate with primary antibody solution for 1 h at 37 °C in a humid chamber (*see* **Note 8**).
8. Wash three times with PBS.
9. Incubate with appropriate secondary antibody (e.g., FITC-labeled anti-mouse-IgG) in a humid chamber for 60 min at 37 °C.
10. Cover the cells with appropriate mounting medium.
11. Analyze samples under a fluorescent microscope according to the properties of your secondary antibody.

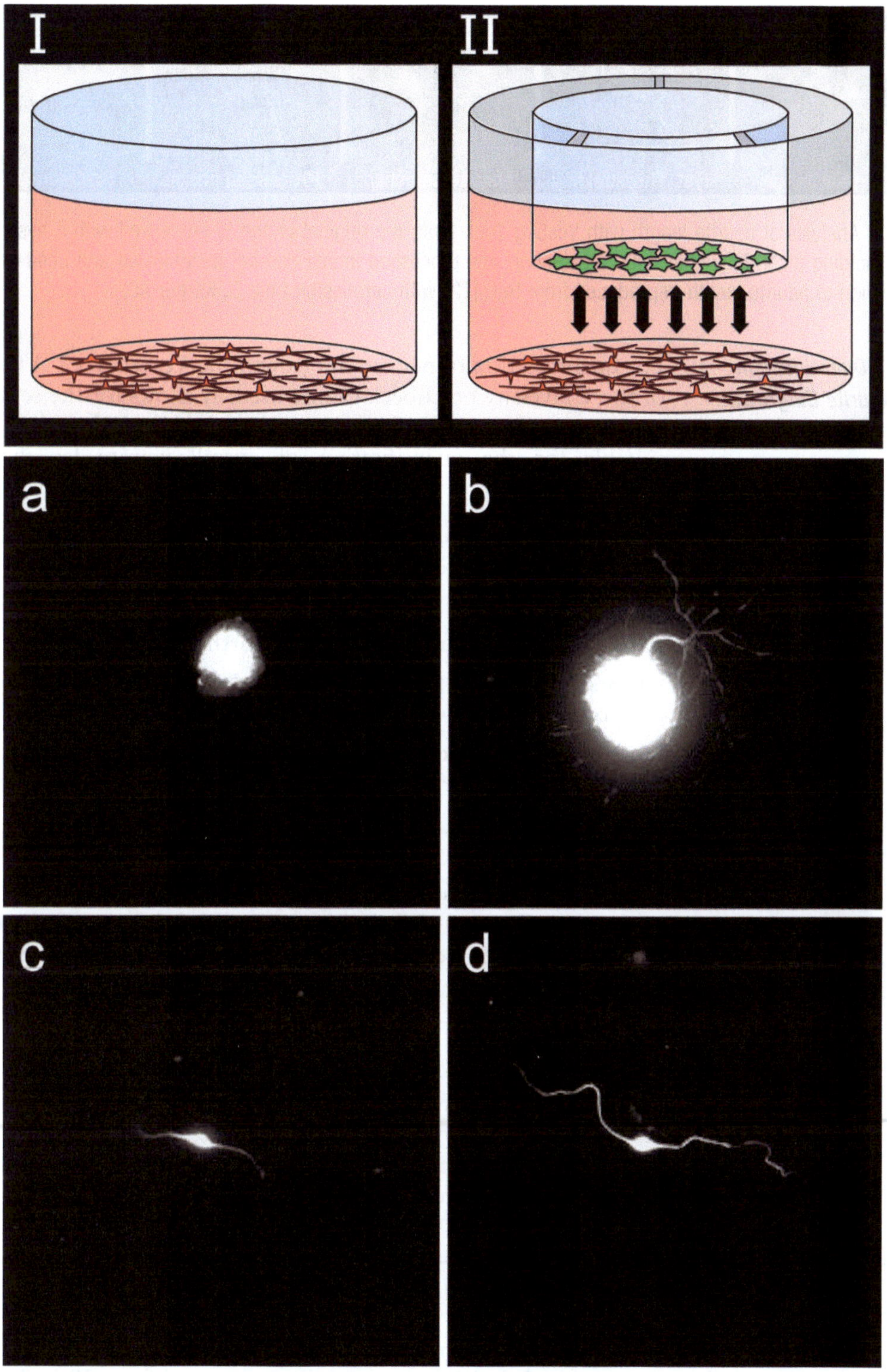

Fig. 1 Setup of the co-culture system. DRGs are grown on coated glass coverslips (**I**) as control. For indirect co-cultures, a preseeded insert with hSGSCs is added on top (**II**). After staining of neuron-specific structures (e.g., neurofilament), this method allows for the comparison of different neurite lengths (**a–d**) (reproduced from Ref. [12] with permission from S. Karger AG)

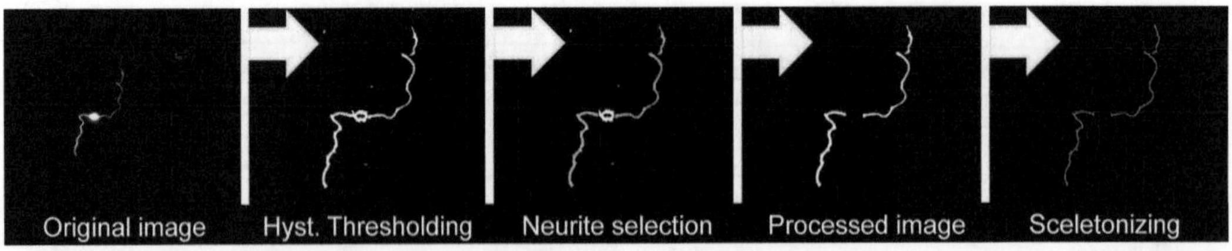

Fig. 2 Analysis of neurite length with VesSeg Tool. First, the original image is processed with a hysteresis thresholding algorithm. After neurite selection, the processed image can be skeletonized and used for the detection of neurite length (reproduced from Ref. [12] with permission from S. Karger AG)

3.9 Quantification of Neurite Outgrowth

For quantification of neurite outgrowth the VesSeg Tool (V0.1.4) of the University of Lübeck can be used. It allows for the semiautomatic segmentation of vascular networks [16–18], but is also suitable for the quantification of overall neurite length (*see* Fig. 2) [12].

1. Download VesSeg Tool V0.1.4 (available at http://www.isip.uni-luebeck.de/?id=150).
2. Open picture.
3. Choose rectangular section to be further analyzed.
4. Start hysteresis thresholding mode using 100% background coverage and 1–4% vessel coverage.
5. Compare the picture with the original one to ensure that all neurites are clearly and continuously present. Otherwise adjust vessel coverage.
6. After calculation change to selection mode.
7. Use right mouse button to deselect non-neurite structures (e.g., cell soma) and the left button to select neurites which have not been detected automatically. Hit space bar and press ok, when done.
8. Choose "Skeletonizing mode" from morphological filters of the image menu (*see* **Note 9**).
9. Read binary image information from image statistic of the image menu.
10. The length given in this table is the neurite length of the processed picture.

4 Notes

1. If you are planning to isolate hSGSCs be sure to generate both media, since FCS concentration is lowered from 20 to 10% in passage 1.

2. hSGSCs can easily be propagated. However, we recommend to use cells up to passage 17.
3. We recommend to use coated coverslips as soon as possible.
4. Even after filtering, you might find a high amount of myelin debris culture dish after seeding the cells. We found that further purification significantly impedes yield and does not improve further analysis, since debris is usually washed off by changing 50% of the culture media after 24 h.
5. Culture plates of some suppliers have notches, which perfectly center matching inserts. If inserts slide to the edge of the culture plate, this may result in a significant variation of cell culture media level. Therefore, we strongly recommend to use matching culture plates from the same supplier with recommended media amounts.
6. For later analysis, neurites must clearly be assigned to individual neurons. Therefore, defined seeding density of DRG neurons must be ensured. We recommend to thoroughly adjust cell density to 4–5 $\times$ 10^4 neurons/cm^2. However, if you experience problems during the analysis, adaption of seeding density may help in the future.
7. In a co-culture setup, medium composition is crucial since hSGSCs and DRG neurons require fundamentally different culture media and supplements. We recommend to include adequate controls to rule out effects which may result solely from the co-culture medium.
8. For reliable staining of neuronal structures, we recommend a mixture of monoclonal antibodies against neurofilament L, M, and H. However, preexisting staining protocols may also be used, if they allow for a robust detection of neuronal structures.
9. Skeletonizing is an important step, since it reduces neurite width to a single pixel. This allows for an accurate calculation of neurite lengths.

References

1. Forda S, Kelly JS (1985) The possible modulation of the development of rat dorsal root ganglion cells by the presence of 5-HT-containing neurons of the brainstem in dissociated cell culture. Brain Res 354(1):55–65
2. Currie KP, Swann K, Galione A, Scott RH (1992) Activation of Ca(2+)-dependent currents in cultured rat dorsal root ganglion neurons by a sperm factor and cyclic ADP-ribose. Mol Biol Cell 3(12):1415–1425
3. Chen JJ, Vasko MR, Wu X, Staeva TP, Baez M, Zgombick JM, Nelson DL (1998) Multiple subtypes of serotonin receptors are expressed in rat sensory neurons in culture. J Pharmacol Exp Ther 287(3):1119–1127
4. Tonge D, Edstrom A, Ekstrom P (1998) Use of explant cultures of peripheral nerves of adult vertebrates to study axonal regeneration in vitro. Prog Neurobiol 54(4):459–480
5. Melli G, Hoke A (2009) Dorsal root ganglia sensory neuronal cultures: a tool for drug discovery for peripheral neuropathies. Expert Opin Drug Discov 4(10):1035–1045. https://doi.org/10.1517/17460440903266829

6. Khammo N, Ogilvie J, Clothier RH (2007) Development of an innervated model of human skin. Altern Lab Anim 35(5):487–491
7. Lebonvallet N, Pennec JP, Le Gall-Ianotto C, Cheret J, Jeanmaire C, Carre JL, Pauly G, Misery L (2014) Activation of primary sensory neurons by the topical application of capsaicin on the epidermis of a re-innervated organotypic human skin model. Exp Dermatol 23 (1):73–75. https://doi.org/10.1111/exd.12294
8. Roggenkamp D, Kopnick S, Stab F, Wenck H, Schmelz M, Neufang G (2013) Epidermal nerve fibers modulate keratinocyte growth via neuropeptide signaling in an innervated skin model. J Invest Dermatol 133(6):1620–1628. https://doi.org/10.1038/jid.2012.464
9. Danner S, Kremer M, Petschnik AE, Nagel S, Zhang Z, Hopfner U, Reckhenrich AK, Weber C, Schenck TL, Becker T, Kruse C, Machens HG, Egana JT (2012) The use of human sweat gland-derived stem cells for enhancing vascularization during dermal regeneration. J Invest Dermatol 132 (6):1707–1716. https://doi.org/10.1038/jid.2012.31
10. Lu CP, Polak L, Rocha AS, Pasolli HA, Chen SC, Sharma N, Blanpain C, Fuchs E (2012) Identification of stem cell populations in sweat glands and ducts reveals roles in homeostasis and wound repair. Cell 150(1):136–150. https://doi.org/10.1016/j.cell.2012.04.045
11. Sebastian A, Volk SW, Halai P, Colthurst J, Paus R, Bayat A (2017) Enhanced neurogenic biomarker expression and reinnervation in human acute skin wounds treated by electrical stimulation. J Invest Dermatol 137 (3):737–747
12. Mehnert JM, Kisch T, Brandenburger M (2014) Co-culture systems of human sweat gland derived stem cells and peripheral nerve cells: an in vitro approach for peripheral nerve regeneration. Cell Physiol Biochem 34 (4):1027–1037. https://doi.org/10.1159/000366318
13. Dietrich PS, McGivern JG, Delgado SG, Koch BD, Eglen RM, Hunter JC, Sangameswaran L (1998) Functional analysis of a voltage-gated sodium channel and its splice variant from rat dorsal root ganglia. J Neurochem 70 (6):2262–2272
14. Hertz DL, Owzar K, Lessans S, Wing C, Jiang C, Kelly WK, Patel J, Halabi S, Furukawa Y, Wheeler HE, Sibley AB, Lassiter C, Weisman L, Watson D, Krens SD, Mulkey F, Renn CL, Small EJ, Febbo PG, Shterev I, Kroetz DL, Friedman PN, Mahoney JF, Carducci MA, Kelley MJ, Nakamura Y, Kubo M, Dorsey SG, Dolan ME, Morris MJ, Ratain MJ, McLeod HL (2016) Pharmacogenetic discovery in CALGB (Alliance) 90401 and mechanistic validation of a VAC14 polymorphism that increases risk of docetaxel-induced neuropathy. Clin Cancer Res 22 (19):4890–4900. https://doi.org/10.1158/1078-0432.CCR-15-2823
15. Wing C, Komatsu M, Delaney SM, Krause M, Wheeler HE, Dolan ME (2017) Application of stem cell derived neuronal cells to evaluate neurotoxic chemotherapy. Stem Cell Res 22:79–88
16. Condurache AP, Mertins A (2012) Segmentation of retinal vessels with a hysteresis binary-classification paradigm. Comput Med Imaging Graph 36(4):325–335. https://doi.org/10.1016/j.compmedimag.2012.02.002
17. Egana JT, Condurache A, Lohmeyer JA, Kremer M, Stockelhuber BM, Lavandero S, Machens HG (2009) Ex vivo method to visualize and quantify vascular networks in native and tissue engineered skin. Langenbeck's Arch Surg 394(2):349–356. https://doi.org/10.1007/s00423-008-0333-3
18. Machens HG, Grzybowski S, Bucsky B, Spanholtz T, Niedworok C, Maichle A, Stockelhuber B, Condurache A, Liu F, Egana JT, Kaun M, Mailander P, Aach T (2006) A technique to detect and to quantify fasciocutaneous blood vessels in small laboratory animals ex vivo. J Surg Res 131(1):91–96

Chapter 12

Engineering a Multilayered Skin Substitute with Keratinocytes, Fibroblasts, Adipose-Derived Stem Cells, and Adipocytes

Maike Keck, Alfred Gugerell, and Johanna Kober

Abstract

A variety of skin substitutes that restore epidermal and dermal structures are currently available on the market. While the main focus in research and clinical application lies in dermal and epidermal substitutes, the development of a subcutaneous replacement, the hypodermis, is often neglected. This chapter describes the use of fibrin sealant as a hydrogel scaffold to generate a three-dimensional skin substitute. For the hypodermal layer adipose-derived stem cells (ASCs) and mature adipocytes are seeded within a fibrin hydrogel. On top, another fibrin clot with incorporated fibroblasts is placed for the construction of the dermal layer. Keratinocytes are added on top of the two-layered construct to form the epidermal layer. The three-layered construct is cultivated for up to 3 weeks with keratinocytes being exposed to air according to the air-liquid interface cultivation model.

Key words Three-layered skin substitute, Keratinocytes, Fibroblasts, Adipose-derived stem cells, Adipocytes, Regenerative medicine, Tissue engineering

1 Introduction

A variety of skin substitutes that restore epidermal and dermal structures are currently available on the market. While the main focus in research and clinical application lies in dermal and epidermal substitutes, the development of a subcutaneous replacement, the hypodermis, is often neglected. Nevertheless, this part of the skin is of major importance as it functions as an insulator, energy storage, and endocrine organ and defines the shape of the body.

This chapter describes the use of fibrin sealant as a hydrogel scaffold to generate a three-dimensional skin substitute. The physical properties of fibrin hydrogel mimic the soft character of fatty tissue and also its mechanical properties resemble those of adipose tissue to a satisfying extent. Therefor fibrin hydrogen is suggested

Sophie Böttcher-Haberzeth and Thomas Biedermann (eds.), *Skin Tissue Engineering: Methods and Protocols*, Methods in Molecular Biology, vol. 1993, https://doi.org/10.1007/978-1-4939-9473-1_12,

to be a suitable candidate for adipose tissue-equivalent formation [1].

For the hypodermal layer adipose-derived stem cells (ASCs) and mature adipocytes are seeded within a fibrin hydrogel. On top, another fibrin clot with incorporated fibroblasts is placed for the construction of the dermal layer. Keratinocytes are added on top of the two-layered construct to form the epidermal layer. The three-layered construct is cultivated for up to 3 weeks with keratinocytes being exposed to air according to the air-liquid interface cultivation model [2].

2 Materials

1. Skin and subcutaneous fat tissue is usually gained from healthy patients undergoing abdominoplasty or other body-contouring surgery (*see* Fig. 1) (*see* **Notes 1** and **2**).

2.1 Culture Media

1. ASC proliferation medium: 10% FCS, 100 units/mL penicillin, 100 μg/mL streptomycin in Dulbecco's modified Eagle medium (DMEM) (*see* **Note 3**).
2. ASCs differentiation medium: Preadipocyte Differentiation Medium (PromoCell GmbH, Heidelberg, Germany) (*see* Table 1).
3. Adipocyte cultivation medium: Adipocyte Nutrition Medium (PromoCell GmbH, Heidelberg, Germany) (*see* Table 1).
4. Fibroblast culture medium: 10% Fetal calf serum, 1% glutamine, 100 units/mL penicillin, 100 μg/mL streptomycin in DMEM.

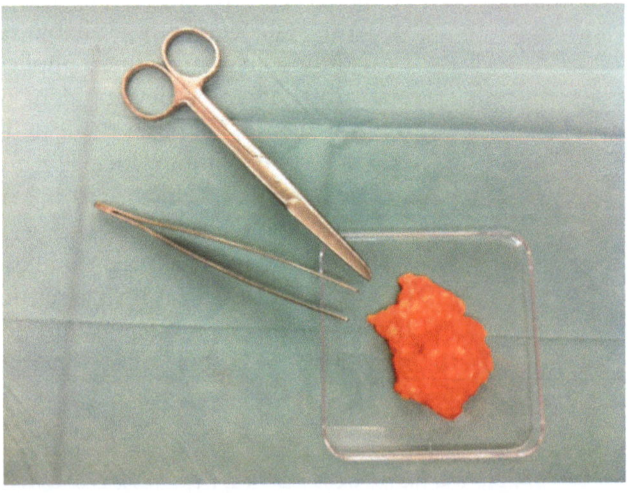

Fig. 1 Starting material: adipose tissue from abdominoplasties

Table 1
Composition of PromoCell media

Final supplement concentrations (after addition to the medium)	Preadipocyte differentiation medium	Adipocyte nutrition medium
Fetal calf serum	–	0.03 mL/mL
d-Biotin	8 μg/mL	8 μg/mL
Insulin (recombinant human)	0.5 μg/mL	0.5 μg/mL
Dexamethasone	400 ng/mL	400 ng/mL
IBMX	44 μg/mL	–
L-Thyroxine	9 ng/mL	–
Ciglitazone	3 μg/mL	–

https://www.promocell.com/products/cell-culture-media/media-for-primary-cells/preadipocyteadipocyte-media/

5. Keratinocyte medium: DermaLife keratinocyte medium.
6. Three-layered skin construct medium: 10% Fetal calf serum in DMEM.

2.2 Reagents

1. Phosphate-buffered saline (PBS).
2. 2% Penicillin/streptomycin.
3. Collagenase solution: 2 mg/mL Collagenase type IV in Hanks' buffered salt solution.
4. AdipoRed staining.
5. 125 mg/mL Fibrinogen.
6. 10 IU/mL Thrombin.
7. Trypsin.

2.3 Equipment

1. 70 μm Cell strainer.
2. Cell culture inserts.

3 Methods

3.1 Isolation of Human Adipose-Derived Stem Cells (ASCs)

1. Wash the adipose tissue in phosphate-buffered saline (PBS) + 2% penicillin/streptomycin (*see* **Note 4**).
2. Mince the adipose tissue with scissors and/or scalpel (*see* Fig. 1). Remove large amounts of connective tissue and vessels.
3. Add the shredded fat tissue into the tubes to collagenase solution (25 mL collagenase + about 10 mL fatty tissue).
4. Incubate in a shaking bath at 37 °C for 1 h.

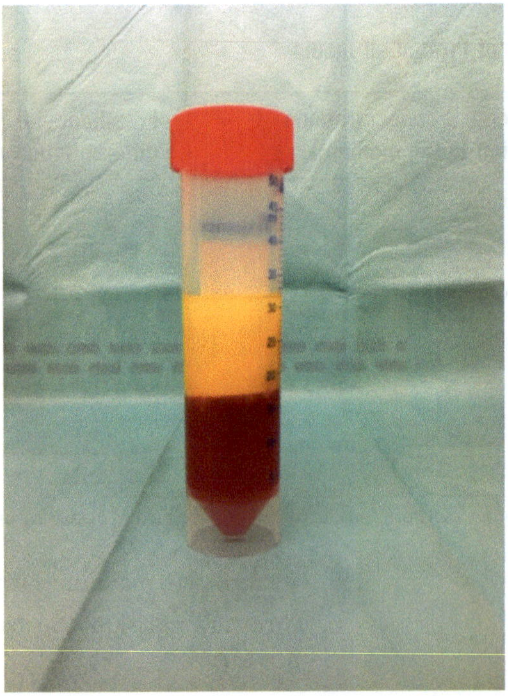

Fig. 2 Different phases after centrifugation: oil (destroyed adipocytes), adipocytes, collagenase, SVF (from top to bottom)

5. Filter the dissolved fatty tissue including collagenase through gauze into a new 50 mL tube.
6. Centrifuge for 10 min at 380 × *g* at 4 ° C.
7. The resulting cell pellet (stromal vascular fraction, SVF) consists of ASCs, macrophages, erythrocytes, endothelial cells, and other cell types (*see* Fig. 2).
8. Carefully aspirate the upper layer including adipocytes (for protocol *see* below).
9. Resuspend pellet with 2 mL red blood cell lysing buffer and incubate for 8 min on ice.
10. Add cold medium and filter the cells through a 70 μm cell strainer.
11. Centrifuge for 7 min at 380 × *g* at 4 ° C.
12. Resuspend pellet with ASC proliferation medium (*see* **Note 5**). ASCs are cultured as a monolayer at 37 °C in ASC proliferation medium in a humidified atmosphere with 5% CO_2. ASC passages up to passage 2 or 3 are recommended for use in skin engineering approaches (*see* Fig. 3).

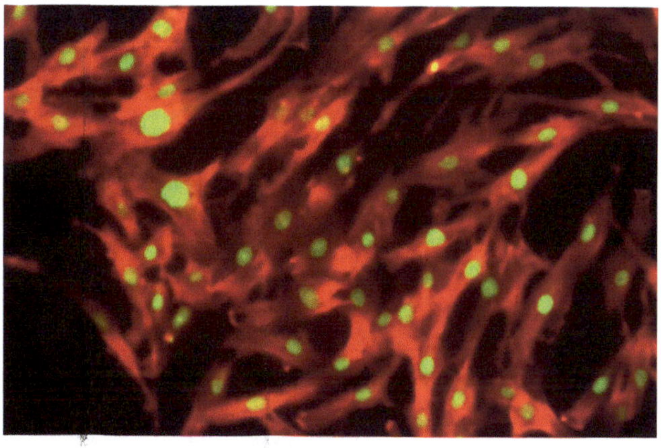

Fig. 3 Isolated ASC. TRITC-phalloidin (cytoskeleton)/SYTOX green (nuclei) staining

Fig. 4 Isolated adipocytes. TRITC-phalloidin (cytoskeleton)/SYTOX green (nuclei) staining

3.2 Isolation of Human Adipocytes

1. Collect freshly isolated adipocytes during the isolation process as described above from the usually discarded adipose phase after the first centrifugation step.
2. Carefully aspirate the upper layer including adipocytes. Possibly start with a 25 mL pipette, as for the upper fat layer the smaller Pasteur pipettes may be too narrow. Culture cells in adipocyte cultivation medium. The cultivation of the cells takes place in incubators at 37 ° C, 5% CO_2, and 95% humidity. Isolated adipocytes are depicted in Fig. 4 (*see* **Notes 6** and 7).

3.3 Adipocyte Differentiation

1. Incubate ASCs for 3 days in differentiation medium.
2. Cells are further incubated in adipocyte cultivation medium. Medium has to be changed every third day.
3. Evaluation of differentiation can histologically be performed by AdipoRed staining (*see* Fig. 5) or by RNA quantification.

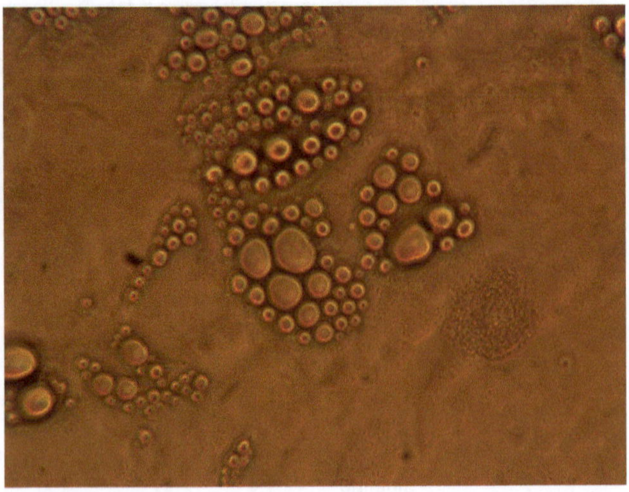

Fig. 5 ASC differentiated into adipogenic lineage (day 24). Intracellular lipid droplets are visible

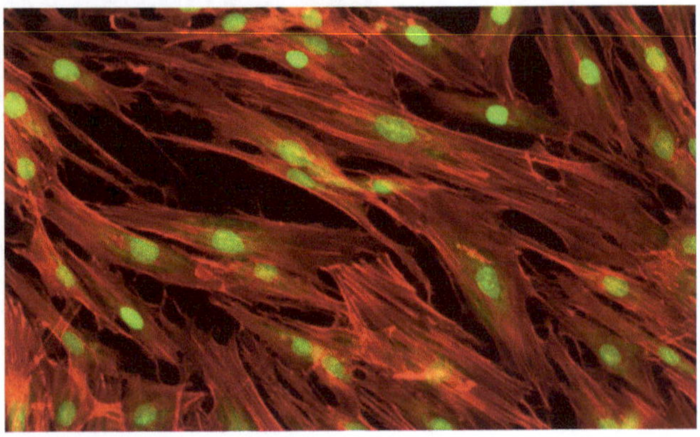

Fig. 6 Isolated fibroblasts. TRITC-phalloidin (cytoskeleton)/SYTOX green (nuclei) staining

3.4 Isolation of Human Dermal Fibroblasts

1. Cut human donor skin into small pieces of about 1 × 1 mm.
2. Skin fragments are put in 6-well plates or dishes and incubated for 14 days in fibroblast culture medium.
3. Skin fragments need to stick to the bottom of the well for fibroblasts to grow out onto the bottom of the plate.
4. Culture upon confluency and split for further cultivation in culture flasks (*see* Fig. 6). The cultivation of the cells takes place in incubators at 37 ° C, 5% CO_2, and 95% humidity.

3.5 Isolation of Human Keratinocytes

1. Cut human skin into pieces of about 2 × 2 cm. Remove subcutis and most of the dermis.
2. Trypsinize for 16 h at 4 °C.

3. Filter the resulting suspension through gauze and centrifuge for 5 min at 1200 rpm.
4. Discard supernatant and resuspend in keratinocyte medium. The cultivation of the cells takes place in incubators at 37 ° C, 5% CO_2, and 95% humidity.

3.6 Construction of a Three-Layered Skin Substitute with a Fibrin Hydrogel Matrix

For fibrin clot formation, equal volumes of fibrinogen and thrombin components have to be mixed in cell culture inserts (*see* **Note 9**).

1. For the hypodermal layer: mix 3×10^5 ASCs with 20 μL of mature adipocytes (about 1.4×10^6 cells/mL) in 100 μL fibrin clots in cell culture inserts (*see* **Note 8**).
2. Incubate clots at 37 °C for 20 min for full polymerization.
3. For the dermal layer: polymerize another fibrin clot on top of the hypodermal layer with 3×10^5 fibroblasts (*see* Fig. 7).
4. Incubate clots at 37 °C for 15 min for full polymerization.
5. For the epidermal layer: add $0.6\text{–}1 \times 10^6$ keratinocytes per clot on top of the dermal layer.
6. The three-layered construct can be cultivated for up to 3 weeks (at 37 °C, 5% CO_2, and 95% humidity) in three-layered skin construct medium where keratinocytes are exposed to air according to the air-liquid interface cultivation model (*see* Fig. 8).

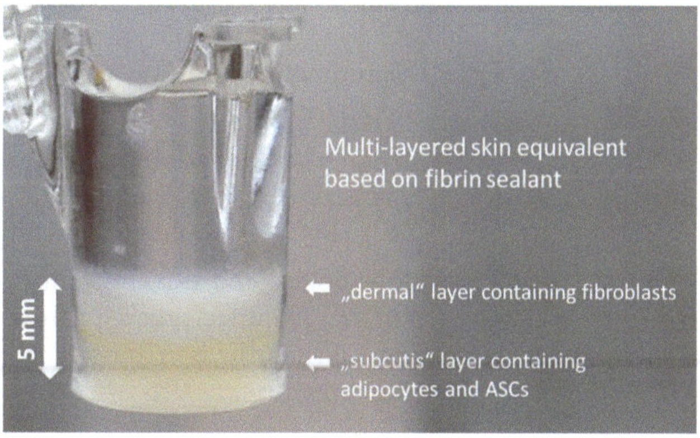

Fig. 7 Two-layered construct

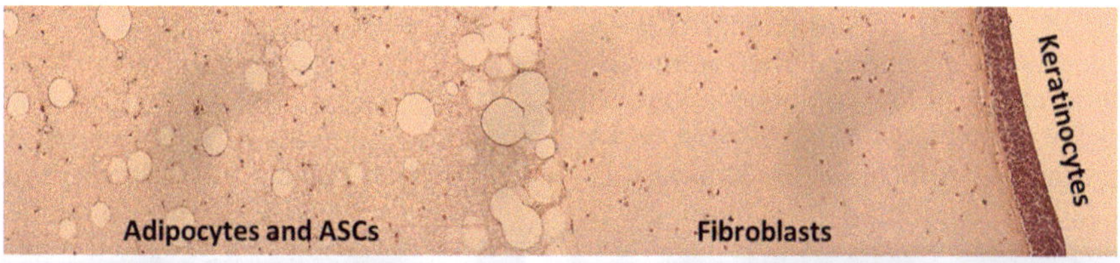

Fig. 8 Three-layered construct

4 Notes

1. There are differences in the viability of cells depending on the donation protocol, especially concerning liposuction methods. As fat tissue is a huge endocrine organ, patients with metabolic disorders like diabetes should be excluded due to eventually altered cell behavior [3].
2. Subjects have to give a written informed consent which has to be approved by the local ethics committee.
3. Required media and solutions are always freshly prepared before use.
4. All cell culture work is done with sterile materials and under sterile workbenches.
5. Isolated cells should be counted in a Bürker-Türk counting chamber with trypan blue stain to prove viability. By cultivation and passaging, the SVF selection and expansion of ASCs will be achieved, and remaining blood cells or endothelial cells will be removed or overgrown in large parts. It is recommended to prove the purity of cells with FACS (e.g., CD90+, CD105+, CD29+, CD44+, CD45−) and also methodically by differentiating in desired lineages (i.e., adipogenic lineage) [4].
6. Adipocytes are sensitive and break easily resulting in an oil film. Adipocytes have to be washed two times with PBS to remove collagenase. Afterwards they can be used for the experiment right away.
7. A long-term cultivation of mature adipocytes is not recommended due to cell break and also possible dedifferentiation [5].
8. In a fibrin clot with high fibrinogen concentration, cells may be stressed due to a high stiffness and density of the hydrogel [6]. High thrombin concentrations may be toxic for cells [7].
9. Fibrinogen and thrombin need to be reconstituted according to the manufacturer's protocol. As fibrinogen/thrombin is used primarily as hemostat or sealer, components should be diluted.

References

1. Peterbauer-Scherb A, Danzer M, Gabriel C, van Griensven M, Redl H, Wolbank S (2012) In vitro adipogenesis of adipose-derived stem cells in 3D fibrin matrix of low component concentration. J Tissue Eng Regen Med 6(6):434–442
2. Kober J, Gugerell A, Schmid M, Kamolz LP, Keck A (2015) Generation of a fibrin based three-layered skin substitute. Biomed Res Int 2015:170427
3. Keck M, Kober J, Riedl O, Kitzinger HB, Wolf S, Stulnig TM et al (2014) Power assisted liposuction to obtain adipose-derived stem cells: impact on viability and differentiation to adipocytes in comparison to manual aspiration. J Plast Reconstr Aesthet Surg 67(1):E1–E8

4. Mildmay-White A, Khan W (2017) Cell surface markers on adipose-derived stem cells: a systematic review. Curr Stem Cell Res Ther 12 (6):484–492
5. Huber B, Kluger PJ (2015) Decelerating mature adipocyte dedifferentiation by media composition. Tissue Eng Part C Methods 21 (12):1237–1245
6. Nürnberger S, Wolbank S, Peterbauer-Scherb A, Morton TJ, Feichtinger GA, Gugerell A et al (2010) Properties and potential alternative applications of fibrin glue. In: von Byern J, Grunwald I (eds) Biological adhesive systems: from nature to technical and medical application. Springer, Vienna
7. Gugerell A, Schossleitner K, Wolbank S, Nurnberger S, Redl H, Gulle H et al (2012) High thrombin concentrations in fibrin sealants induce apoptosis in human keratinocytes. J Biomed Mater Res A 100a(5):1239–1247

Chapter 13

Fabrication of Chimeric Hair Follicles for Skin Tissue Engineering

Andrea L. Lalley and Steven T. Boyce

Abstract

Fabrication of engineered skin substitutes provides an alternative approach for the treatment of full-thickness burns and other skin injuries. Improving the functionality of current skin substitute models requires incorporation of skin appendages, including hair follicles, sebaceous glands, and sweat glands. In this chapter, methods for generating skin substitutes incorporating chimeric hair follicles are described. Isolation of human keratinocytes, human fibroblasts, and murine dermal papilla cells is first outlined. These cell types are then combined with collagen-glycosaminoglycan (GAG) scaffolds to generate human-murine chimeric grafts which are then grafted to full-thickness surgical wounds in immunodeficient mice. The methods described allow for the generation of a human-mouse follicular structure.

Key words Engineered skin, Hair follicles, Human keratinocytes, Murine dermal papilla cells, Chimeric skin

1 Introduction

The morphogenic processes leading to human hair follicle formation result from a regulated series of mesenchymal-epithelial interactions occurring between 10 and 14 weeks of human embryonic development [1, 2]. Interestingly, these processes are likely exclusive to fetal development and are not reactivated at postnatal time points. While fundamental components of the developmental mechanisms have been well defined [3, 4], further investigation is required to more fully elucidate strategies to generate human hair follicles in vivo.

Identifying a suitable dermal cell source that is capable of producing signals to guide neogenesis of pilosebaceous units remains an unresolved consideration in the field. The use of human fetal dermal cells represents one potential approach toward resolving this issue. Wu et al. successfully generated entirely human hair follicles by combining dermal cells derived from human fetal

Sophie Böttcher-Haberzeth and Thomas Biedermann (eds.), *Skin Tissue Engineering: Methods and Protocols*, Methods in Molecular Biology, vol. 1993, https://doi.org/10.1007/978-1-4939-9473-1_13,

scalp tissue [estimated gestational age 15–18 weeks] with neonatal foreskin keratinocytes following engraftment to immunodeficient mice [5]. However, this approach is seldom implemented as the use of human fetal cells derived from elective abortions is prohibited by most states' laws in the USA. The use of adult human dermal cells, specifically human dermal papilla cells (hDPCs), has also been investigated as a method to recapitulate aspects of hair follicle generation; however, the use of this cell type has been met with limited success due to the finite potency of hDPCs [6, 7].

Recently, our laboratory group has developed a chimeric model of hair follicle formation using a combination of culture-expanded dermal papilla cells derived from mouse vibrissae (mDPCs) and human neonatal foreskin keratinocytes [8–10]. This combination resulted in the formation of chimeric follicular structures and pigmented hairs in our engineered skin substitute (ESS) model. This chapter describes how to generate murine-human chimeric grafts by first describing methods for murine and human cell isolation and culture, followed by methods for generating the chimeric graft, and transplantation procedures in animal models.

2 Materials

2.1 Mice and Cell Sources

1. Human neonatal foreskin—average size: 1–2 cm^2 [11].
2. C57BL/6-Tg(UBC-GFP)30 Mouse (Jackson Laboratory, 004353).
3. NIH-III Nude Homozygous Mouse [NIH-$Lyst^{bg\text{-}J}$*Foxn1*-^{nu}*Btk*^{xid}] (Charles River Laboratory, Strain Code: 201).

2.2 Cell Isolation and Culture

1. HEPES buffered saline (HBS) [12].
2. Modified MCDB153 (mMCDB153) medium + supplements [12], *see* Table 1.
3. Low-calcium mMCDB153 medium + supplements [12], *see* Table 1.
4. AmnioMAX™-C100 medium+ supplement provided with kit (Life Technologies), *see* Table 1.
5. DMEM, −phenol red, sodium bicarbonate medium + supplements [12], *see* Table 1.
6. 2.4 U/mL Dispase in HBS.
7. 0.025% Trypsin + 0.01% EDTA.
8. 625 U/mL Collagenase.
9. Bovine pituitary extract (BPE).
10. Fetal bovine serum (FBS).

Table 1
Supplements for selective culture of cells for preparation of chimeric grafts

	Human fibroblasts	Human keratinocytes	Murine dermal papilla cells
Basal medium	DMEM [12]	mMCDB153 [12]	AmnioMAX™-C100[a]
Epidermal growth factor	10 ng/mL	1 ng/mL	–
Insulin	5.0 μg/mL	5.0 μg/mL	–
Hydrocortisone	0.50 μg/mL	0.50 μg/mL	–
Bovine pituitary extract	–	0.2%v/v	–
Fetal bovine serum	4%v/v	–	–
Penicillin-streptomycin-fungizone	1%v/v	1%v/v	1%v/v

[a]Supplement AmnioMAX™-C100 basal media with supplement bottle provided in the kit

11. Coating matrix kit protein.
12. Dettol: 5% v/v Milli-Q H_2O, sterile filtered.
13. Stainless steel forceps, high precision, blunt and fine.
14. Dissecting microscope.
15. 50 mL Conical tubes.
16. Scalpel handle and blades, surgical scissors, blunt forceps, razor blade.
17. 70% Ethanol.
18. Hemostat.
19. T-25, T-150, T-225 tissue culture flasks; 100 mm tissue culture dishes; 6-well dish.
20. 70 μm Cell strainer.
21. Steriflip 50 mL disposable filter units.
22. Ziplock bags.

2.3 Chimeric Graft Generation

1. Collagen-GAG scaffold (produced in-house) [13, 14].
2. Lifting frame (custom made) [12].
3. UCMC160 medium + supplements [11], *see* Table 1.
4. Sterile cotton pads (cut into 8 cm × 8 cm squares).
5. 150 mm Tissue culture dishes
6. Merocel, 13 mm diameter (Medtronic).

2.4 Grafting Surgical Procedure

1. N-Terface® dressing.
2. Insulin syringes U-100.
3. 1 mL TB syringe.

4. Sterile drapes.
5. Alcohol pads.
6. Povidone-Iodine Prep Pads.
7. Plain gauze—surgical.
8. Sutures—Ethilon® nylon, 18″, 6–0.
9. Avertin (prepared in-house) [12, 15].
10. Tegaderm™ transparent film dressing.
11. Coban™ self-adhering wrap.
12. Steri-strip compound, benzoin tincture.
13. Antibiotic ointment (prepared in-house).
14. Tongue depressor, sterile.
15. Surgical marker.
16. Sharp scissors, needle driver.
17. Saline.
18. Earhole punch.
19. Ziplock bags.
20. Sterile surgical gloves.
21. 2 cm × 2 cm Cardboard square.
22. Heating pads.

3 Methods

All procedures are performed by aseptic technique using a biological safety cabinet.

Place instruments into 250 mL beaker containing 100 mL of 95% ethanol and sterilize any instruments used in this procedure by flaming/cooling immediately prior to use.

Store cells/reagents on ice throughout the procedures.

Turn on refrigerated centrifuge prior to initiating any procedures.

Figure 1 provides an overview of the procedures detailed in this section.

3.1 Isolation and Cell Culture of Human Neonatal Foreskin Keratinocytes (hK) and Fibroblasts (hF)

1. General preparation procedures:
   (a) Thaw Dispase solution, and warm supplemented DME, supplemented mMCDB 153, and HBS in a 37 °C water bath for ≥30 min.
   (b) For tissue decontamination, prepare 5% (vol/vol) Dettol solution. Add 1.0 mL Dettol to 19.0 mL Milli-Q H_2O and sterile filter using the Steriflip disposable filter unit (day 1 only).

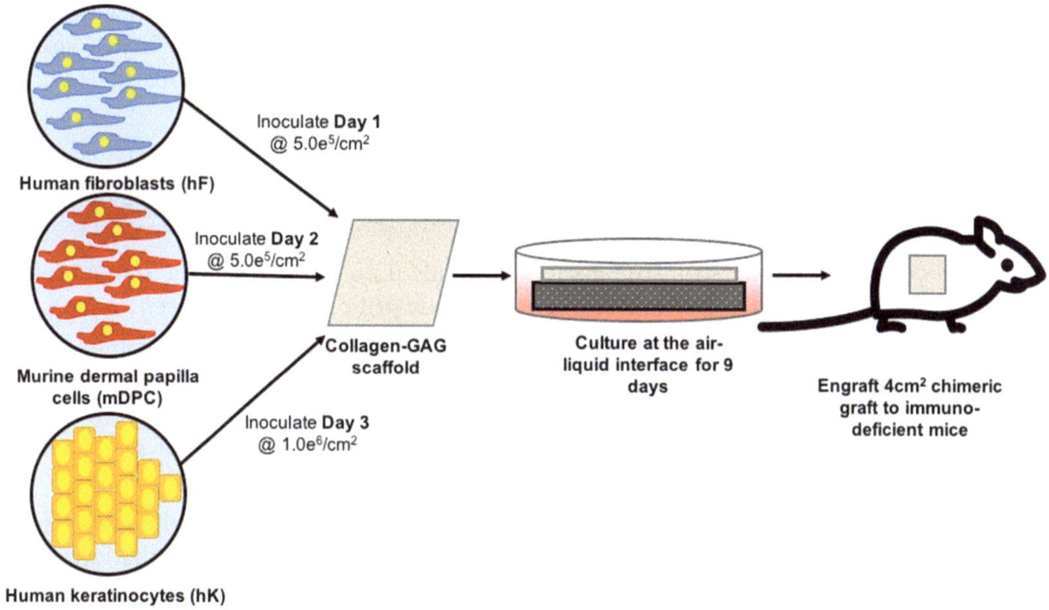

Fig. 1 Preparation of chimeric ESS from human and murine cell sources. Cells are grown in culture, inoculated on a collagen-GAG scaffold, cultured at the air-liquid interface in maturation medium for 9 days, and grafted to full-thickness wounds in immuno-deficient mice

(c) Warm HBS, FBS, supplemented mMCDB153, and trypsin-EDTA in a 37 °C water bath for ≥30 min (day 2 only).

(d) For trypsin-EDTA deactivation, prepare 100 mL of supplemented mMCDB153 + 10% FBS. Aliquot 20 mL into each of the two 50 mL conical tubes and store on ice until ready to use. Store the remaining volume on ice for later use (day 2 only).

2. Decontamination and tissue preparation (day 1):

   (a) Acquire 3–5 human neonatal foreskin tissue samples.

   (b) Place foreskins in supplemented mMCDB153 at 4 °C until the time of cell isolation. Isolation should be completed on the day of foreskin collection to maximize cell yield and viability.

   (c) Wash the tissue 2× with 20 mL HBS for 20 s in a 100 mm tissue culture dishes to remove residual blood products.

   (d) Transfer tissue (using sterile blunt forceps) to a 100 mm tissue culture dish containing 10 mL 5% Dettol solution. Wash for 15–30 s, ensuring that all surfaces are exposed to the Dettol solution.

(e) Transfer tissue (using sterile blunt forceps) to a 100 mm tissue culture dish containing 20 mL HBS. Gently agitate the tissue with the blunt forceps to ensure that Dettol solution is rinsed from the tissue.

(f) Repeat HBS rinse 2×.

(g) Record the size, shape, and general appearance of the tissue samples.

(h) Cut tissue into strips 0.25 cm wide using sterile scalpel and sterile razor blade gripped with hemostat.

(i) Transfer cut tissue strips into the Dispase solution in a 50 mL conical tube using sterile blunt forceps, allowing tissue strips to become submerged in the solution. Ensure a ratio of Dispase to tissue sample area of no less than 1.5 mL/cm^2.

(j) Place the 50 mL conical tube containing the tissue strips into a refrigerator at 4 °C for overnight incubation.

3. Tissue preparation and cell isolation (day 2):

(a) Prepare 3–4 T225 culture flasks for hK inoculation (dependent on the number and areas of foreskins received).

- Add 15 mL of dilution medium from the Cascade Biologics coating matrix kit to each flask and rock back and forth to ensure uniform distribution.
- Add 150 μL of coating matrix from the Cascade Biologics coating matrix kit to each flask and rock back and forth to ensure uniform distribution.
- Cap the flasks and incubate at room temperature for 30 min.
- Following incubation, remove excess coating matrix/dilution medium from each flask.
- Add 40 mL of warmed supplemented mMCDB153 to each flask and place in incubator for ≥30 min to equilibrate prior to cell inoculation.

(b) Set up three 100 mm dishes, each containing 20 mL of warmed HBS. Label one dish with a "T," the second dish with an "E," and the third dish with a "D."

(c) Remove the dish from the refrigerator containing foreskin tissue strips and transfer the tissue strips to 100 mm dish labeled "T" containing 20 mL HBS.

(d) Using sterile blunt forceps, separate the dermis from the epidermis, placing the epidermal strips in the dish labeled "E" and the dermal strips in the dish labeled "D."

4. Human foreskin keratinocyte isolation from epidermis:
   (a) Set up the following 50 mL conical tubes in preparation for epidermal cell isolation: Tube 1—20 mL 0.025% trypsin + 0.01% EDTA, tube 2—20 mL supplemented mMCDB153 + 10% FBS, tube 3—20 mL HBS, tube 4—20 mL supplemented mMCDB153 + 10% FBS, tube 5—20 mL HBS, and tube 6—20 mL supplemented mMCDB153 + 10% FBS.
   (b) Transfer epidermal strips using sterile blunt forceps into tube 1 (containing 0.025% trypsin + 0.01% EDTA solution (1 mL/cm^2 of tissue area)). Use a 25 mL pipet coated with supplemented mMCDB 153 + 10% FBS to pipet/agitate the epidermal strips for 3–5 min in the 0.025% trypsin + 0.01% EDTA solution.
   (c) Filter the cell suspension through a 70 μm cell strainer into a 50 mL tube containing supplemented mMCDB153 + 10% FBS (#2).
   (d) Transfer the epidermal strips to a 50 mL tube (#3) containing 20 mL HBS. Repeat agitation with a 25 mL pipet coated with supplemented mMCDB 153 + 10% FBS to pipet/agitate the epidermal strips for 3–4 min.
   (e) Filter the cell suspension through a 70 μm cell strainer into a 50 mL tube containing 20 mL supplemented mMCDB153 + 10% FBS (#4).
   (f) Transfer the epidermal strips to a 50 mL tube (#5) containing 20 mL HBS. Repeat agitation with a 25 mL pipet coated with supplemented mMCDB 153 + 10% FBS to pipet/agitate the epidermal strips for 3–4 min.
   (g) Filter the cell suspension through a 70 μm cell strainer into a 50 mL tube containing 20 mL supplemented mMCDB153 + 10% FBS (#6).
   (h) Centrifuge all cell suspensions for 7 min at 220 g (4 °C). Resuspend the cell pellets in cold supplemented mMCDB153 and calculate the total cell number.
   (i) Inoculate cells into equilibrated T225 flasks at a cell density of 2e^3 keratinocytes/cm^2 for a 5-day culture period.
   (j) *At 24 h postinoculation*: Aspirate culture medium and replace with warmed, supplemented mMCDB153 containing 0.06 mM (low) calcium. Change the medium using low-calcium mMCDB153 medium throughout the P0 culture period.
   (k) When cells reach 75–90% confluency, either cryopreserve the cultured cells or expand cultured cells to generate sufficient cell numbers for generation of chimeric grafts.

(l) Subculture hK to passage 2 (using mMCDB153; containing 0.2 mM calcium; and, discontinue the coating matrix) for generation of chimeric graft.

5. Human foreskin fibroblast isolation from dermis:

(a) Dissolve collagenase powder into supplemented 153 + 5%(v/v) BPE to produce a final concentration of 625 U/mL. To do so, combine 28.5 mL supplemented mMCDB153 and 1.5 mL BPE into a 50 mL conical tube. Pipet small amounts (2–3 mL) into the collagenase vial until all collagenase is dissolved. Filter sterilize this solution using a 50 mL conical tube steriflip and place on ice until the time of use.

(b) Mince the dermal tissue by first grasping a sterile razor blade firmly with a hemostat and using a scalpel blade to mince the tissue into 0.5–1.0 mm^3 pieces.

(c) Collect the minced dermal pieces and transfer to the sterile 50 mL conical tube containing collagenase. Maintain a ratio of the collagenase solution to biopsy area of 1.0 mL/cm^2.

(d) Agitate with a 25 mL pipet for 1 min. Incubate the solution at 37 °C for 20–30 min. Agitate the dermis every 10 min for 1–2 min. Collagenase incubation is sufficient when the dermal pieces develop filamentous fragments and begin shedding small fragments. Once this point is reached, end the incubation by centrifuging for 7 min at 220 × *g* (4 °C).

(e) Resuspend the pellet in 30 mL supplemented DME and incubate for 5–10 min at room temperature.

(f) Centrifuge for 5 min at 220 × *g* (4 °C). Resuspend tissue pieces in 2 mL of supplemented DME for each T150 to be inoculated. Place inoculated flasks into incubator.

(g) *At 24 h postinoculation*: Set up an equal number of T-150 tissue culture flasks used to inoculate the dermal pieces on day 1. Add 10 mL supplemented DME to each flask and place in incubator to equilibrate for ≥30 min.

(h) Collect media from all flasks inoculated on day 1 in a 50 mL conical tube. Centrifuge for 5 min at 220 × *g* (4 °C).

(i) Aspirate supernatant and resuspend pieces in 5 mL of supplemented DME. Inoculate pieces equally into T-150 tissue culture flasks and return to incubator.

(j) *At 48 h postinoculation*: Set up half the number of T-150s used to initially inoculate fibroblasts. Add 10 mL of supplemented DME to each flask and place in incubator to equilibrate for ≥30 min.

(k) Collect media from all flasks inoculated on day 1 in a 50 mL conical tube. Centrifuge for 5 min at 220 × *g* (4 °C).

(l) Aspirate supernatant and resuspend pieces in 5 mL of supplemented DME. Inoculate pieces equally into T-150 tissue culture flasks and return to incubator.

(m) When cells reach 75–90% confluency, either cryopreserve the cultured cells (according to a traditional cryopreservation protocol) or subculture the cells to generate sufficient cell numbers for generation of chimeric grafts.

(n) Subculture hF to passage 2 for generation of chimeric grafts.

3.2 Isolation of Murine Dermal Papilla Cells (mDPCs)

1. General preparation procedures:
   (a) Warm supplemented AmnioMax™-C100 and HBS in 37 °C water bath for ≥30 min.
   (b) Transfer dissecting microscope to laminar flow hood for murine dermal papilla dissection.
2. Decontamination and tissue preparation:
   (a) Prepare 2–3 T-25 culture vessels for murine dermal papilla culture.
      - Add 1 mL of dilution medium from the Cascade Biologics coating matrix kit to each flask and rock back and forth to ensure uniform distribution.
      - Add 10 μL of coating matrix from the Cascade Biologics coating matrix kit to each flask and rock back and forth to ensure uniform distribution.
      - Cap the flasks and incubate at room temperature for 30 min.
      - Following incubation, remove excess coating matrix/dilution medium from each flask.
      - Add 2–3 mL of warmed supplemented AmnioMAX™-C100 to each flask and place in incubator for ≥30 min to equilibrate prior to mDP inoculation.

   (b) Acquire adult male C57BL/6-Tg(UBC-GFP)30 mouse.
   (c) Euthanize mouse according to prevailing regulatory requirements for the humane use of animals in research. Transfer euthanized mouse to laminar flow hood. Isolation should be completed on the day of euthanization to maximize cell yield and viability.
   (d) Place the mouse in a 100 mm tissue culture dish. Douse the mouse with 70% ethanol, specifically in the region of the vibrissal pads.

(e) Using tissue forceps, grip the proximal edge of the vibrissal pad (near the edge of vibrissae growth). Use a scalpel to create an incision along this edge. Pull up on the vibrissal pad with the tissue forceps and continue the incision along the nasal bone until the nose is reached. Repeat this procedure on the contralateral side. Place the vibrissal pads in HBS + 10% antibiotic/antimycotic in a 100 mm tissue culture dish. Rinse the vibrissal pads with HBS + 10% antibiotic/antimycotic two times.

(f) Place the lid from a 100 mm tissue culture dish (upside down) under the dissecting microscope to serve as a sterile dissection surface. Remove the vibrissal pad from the HBS + 10% antibiotic/antimycotic solution and place on the dissection surface.

(g) Pipet 300 μL droplets of HBS + 10% antibiotic/antimycotic along the outside edges of the dissection surface.

(h) Begin by isolating individual vibrissal follicles. To do so, use sterile fine-tip forceps and secure an edge of the tissue piece for leverage. Identify an intact follicle in anagen phase and, using another fine-tip forceps, gently grasp the follicle approximately halfway along its length (where the vibrissa shaft exits at the skin surface). Pull gently on the follicle and the follicle should become isolated from the surrounding tissue and fascia. Place the follicle into one of the droplets along the outer edge of the dissection surface. Continue this process until all follicles have been isolated from both vibrissal pads.

(i) Place a new lid from a 100 mm tissue culture dish (upside down) under the dissecting microscope and pipet 300 μL volumes of HBS + 1% antibiotic/antimycotic along the outside edges of the dissection surface. Transfer the follicles from the first dish to the droplets in the second dish.

(j) Under the microscope, isolate one follicle and using a scalpel transect the follicle through the matrix just proximal to the dermal papilla. Using a fine-tip forceps in one's left hand, pinch down on the left edge of the opened end bulb to hold the bulb in place (right-handed operator). Use an 18 gauge needle in one's right hand and gently push up on the bottom of the end bulb to expose the dermal papilla. Use a second 18 gauge needle to isolate the dermal papilla from the surrounding matrix.

(k) Transfer the dermal papilla to a collagen-coated T-25 tissue culture flask using a 200 μL micropipet. Ensure that there is an appropriate level of medium so that the dermal papilla attaches to the bottom of the dish and is not floating. Repeat these steps until all of the dermal papillae have been isolated from the follicles. Inoculate 5–7 dermal papillae into each T-25 tissue culture flask.

3. Dermal papilla cell culture:
   (a) Monitor explant outgrowth following inoculation. Continue to maintain adequate medium levels by adding supplemented AmnioMAX™-C100 as needed. mDPCs should begin migrating from dermal papilla 3–5 days postexplant. Once cells are present, change AmnioMAX™-C100 every 2 days.
   (b) Subculture at 2–3 × 10^3 cells/cm^2 into T-225 tissue culture flasks. Cells should reach confluency within 5–6 days.

3.3 Preparation of Chimeric Graft

1. Rehydration and preparation of Merocel:
   (a) *24 h prior* to graft generation:
   - Warm unsupplemented UCMC160 in 37 °C water bath for ≥30 min prior to use.
   - Place a pair of blunt forceps in 100 mL of 95% ethanol in a 250 mL beaker. Sterilize instruments prior to use by passing through a flame and allowing ethanol to burn off. Allow instrument to cool prior to use.
   - Prepare 1–3 150 mm cell culture dishes by adding 120 mL of warmed unsupplemented UCMC160 to each dish.
   - Open peel pack containing sterile Merocel and using sterile blunt forceps transfer the Merocel to 150 mm dish containing UCMC160.
   - Place dishes containing rehydrated Merocel to 37 °C incubator overnight.

   (b) Day of graft inoculation:
   - Retrieve rehydrated Merocel from incubator and aspirate UCMC160 from each dish. Ensure that the medium is removed by tilting dish and using an aspirating pipet to remove excess medium from the Merocel sponge.
   - Refill each dish with 60 mL of supplemented UCMC160 and return dishes to 37 °C incubator for 60 min, or until needed.

2. Rehydration and preparation of collagen-GAG scaffold [*same day of graft preparation*]:
   (a) Warm HBS in a 37 °C water bath for ≥30 min prior to use.
   (b) Prepare 1–3 150 mm culture dishes by dispensing 25 mL of warmed HBS into each dish.
   (c) Place a 9.5 cm × 9.5 cm piece of N-Terface into each dish using sterile forceps. Ensure that N-Terface is completely wet.
   (d) Transfer one sterile collagen-GAG scaffold to each dish, positioning over the N-Terface. Allow to incubate at room temperature for 30 min.
   (e) Aspirate HBS and add 25 mL of supplemented UCMC160. Allow to incubate at room temperature for 30 min.
   (f) Aspirate supplemented UCMC160 and add 25 mL of supplemented UCMC160. Place in 37 °C incubator to equilibrate for a minimum of 30 min and continue to store in incubator until ready for use.
   (g) Prepare 1–3 150 mm dishes by placing one 8 cm × 8 cm piece of sterile cotton pads into each dish. Add ~10 mL of supplemented UCMC160 to hydrate the cotton pad.
   (h) Retrieve 150 mm dishes containing the hydrated collagen-GAG scaffold. Center the collagen-GAG scaffold over the N-Terface and transfer to the hydrated sterile cotton piece.
   (i) Using a sterile scalpel blade and blunt forceps, cut the collagen-GAG scaffold (with N-Terface) into 2 cm × 2 cm squares.
   (j) Retrieve 150 mm dishes containing both the rehydrated Merocel and the collagen-GAG scaffold. Aspirate medium from Merocel. Transfer collagen-GAG scaffold with N-Terface to Merocel (N-Terface should be on the underside of the collagen, in between the Merocel and collagen-GAG scaffold). Ensure that there are no wrinkles or bubbles in the collagen-GAG scaffold. Return Merocel with collagen-GAG scaffold to the incubator until the time of cell inoculation.
3. Preparation of lifting frame [*same day of graft generation*]:
   (a) Warm supplemented UCMC160 in a 37°C water bath for ≥30 min prior to use.
   (b) Prepare 1–3 150 mm dishes by opening sterilized lifting frame packets and place lifting frames into the dishes. Place sterile cotton pad on top of lifting frame.

(c) Dispense supplemented UCMC160 on top of cotton to hydrate. Once hydrated, continue dispensing supplemented UCMC160 under the lifting frame until medium level reaches through the meshed surface and contacts the underside of the cotton pad.

(d) Return lifting frame dishes to the incubator until needed.

4. General preparation procedures [*same day of graft generation*]:

   (a) Thaw 0.025% trypsin + 0.01% EDTA solution (for keratinocyte harvest—5 mL/T225). Warm supplemented mMCDB 153, FBS, UCMC160, and HBS in warm water bath for $\geq$30 min.

5. Human fibroblast harvest and inoculation [day 1]:

   (a) Harvest hF per standard protocols. Resuspend passage 3 hF in supplemented UCMC160 and count cells with a standard hemocytometer.

   (b) The target inoculum is 5e5 hF/cm^2. Based on total cell numbers, determine how many chimeric grafts will be generated and calculate the required numbers. Centrifuge the harvested fibroblasts for 7 min at 220 × *g* (4 °C).

   (c) Prepare the cell inoculum as follows:

   - Multiply the area of the collagen-GAG scaffold by 0.167 mL/cm^2 to determine the cell inoculum volume.
   - Aspirate the supernatant from the cell pellet and resuspend in UCMC160 to a concentration of 3.0×10^6 fibroblasts/mL.
   - Retrieve dishes with lifted collagen-GAG scaffolds and place in hood. Aspirate medium from the dish. Using a serological pipet, pipet half of the inoculation volume dropwise in one direction. Rotate the dish by 90° and pipet the remaining inoculum volume. Ensure even distribution of inoculum volume.
   - Retrieve dishes containing lifting frames from the incubator.
   - Allow the inoculum volume to soak through for no longer than 60 min. Once the inoculum volume has soaked through (usually 10–20 min), use sterile blunt forceps and transfer the collagen-GAG scaffold to the dishes containing lifting frames. Return to incubator.

   (d) Repeat **steps 3.3.1(a)–3.3.5(c)** for any additional scaffolds.

   (e) Add 60 mL of UCMC160 to dishes containing Merocel and return to incubator for cell inoculations on days 2 and 3.

6. Murine DPC harvest and inoculation [day 2]:
   - (a) Harvest mDPCs per standard protocols. Resuspend passage 3 mDPCs in supplemented UCMC160 and count cells according to standard protocol.
   - (b) The target inoculum is 5e5 mDPC/cm^2.
   - (c) Prepare the cell inoculum as follows:
     - Multiply the area of the collagen-GAG scaffold by 0.167 mL/cm^2 to determine the cell inoculum volume. Note: The scaffold may have decreased in area due to the presence of the previously inoculated fibroblasts.
     - Aspirate the supernatant from the cell pellet and resuspend in UCMC160 to a concentration of 3.0×10^6 mDPCs/mL.
     - Retrieve dishes with lifted collagen-GAG scaffolds and dishes with Merocel into the hood. Aspirate medium from the Merocel dish and transfer the collagen-GAG scaffold from the lifting frame to the Merocel.
     - Using a serological pipet, pipet half of the inoculation volume dropwise in one direction. Rotate the dish by 90° and pipet the remaining inoculum volume. Ensure even distribution of inoculum volume.
     - Retrieve dishes containing lifting frames from the incubator.
     - Allow the inoculum volume to soak through for no longer than 60 min. Once the inoculum volume has soaked through (usually 10–20 min), use sterile blunt forceps and transfer the collagen-GAG scaffold to the dishes containing lifting frames. Return to incubator.
   - (d) Repeat **steps 3.3.6(a)–3.3.6(c)** for any additional scaffolds.
   - (e) Add 60 mL of UCMC160 to dishes containing Merocel and return to incubator for cell inoculation on day 3.
7. Human keratinocyte harvest and inoculation [day 3]:
   - (a) Harvest hK per standard protocols. Resuspend passage 3 hK in supplemented UCMC160 and count cells according to standard protocol.
   - (b) The target inoculum is 1e6 hK/cm^2.
   - (c) Prepare the cell inoculum as follows:
     - Multiply the area of the collagen-GAG scaffold by 0.083 mL/cm^2 to determine the cell inoculum volume. Note: The scaffold may have decreased in area due to the presence of the previously inoculated fibroblasts.

- Aspirate the supernatant from the cell pellet and resuspend in UCMC160 to a concentration of 1.2×10^7 mDPCs/mL.
- Retrieve dishes with lifted collagen-GAG scaffolds and dishes with Merocel into the hood. Aspirate medium from the Merocel dish and transfer the collagen-GAG scaffold from the lifting frame to the Merocel.
- Using a serological pipet, pipet half of the inoculation volume dropwise in one direction. Rotate the dish by 90° and pipet the remaining inoculum volume. Ensure even distribution of inoculum volume.
- Allow the inoculum volume to soak through for no longer than 60 min. Once the inoculum volume has soaked through (usually 30–40 min), use sterile blunt forceps and transfer the collagen-GAG scaffold to the dishes containing lifting frames. Return to incubator.

(d) Repeat **steps 3.3.7(a)–3.3.7(c)** for any additional scaffolds.

8. Postinoculation procedures:
   (a) Maintain grafts on lifting frames at the air-liquid interface for 9 days.
   (b) Medium (UCMC160) should be changed daily.
   (c) Collect biopsy samples for histological analysis, and/or collect evaluations of epidermal barrier [9] as needed.

3.4 Chimeric Grafting Procedure

1. Graft preparation for transport:
   (a) Prepare 1–3 150 mm dishes by placing one 8 cm × 8 cm piece of sterile cotton pad into each dish. Add ~5 mL of supplemented UCMC160 to hydrate the cotton pad. Ensure that cotton is fully hydrated, but there is no excess medium in the dish.
   (b) Using a pair of sterilized sharp scissors and blunt forceps, cut N-Terface into 2 cm × 2 cm pieces for each of the chimeric grafts and store in a 150 mm dish containing 25 mL of supplemented UCMC160.
   (c) Retrieve lifting frames with chimeric grafts from the incubator. Using sterile blunt forceps, transfer chimeric grafts with N-Terface to 150 mm dish containing hydrated cotton. Place up to four 2 cm × 2 cm grafts on each piece of cotton for transport.
   (d) Once on the cotton pad, place an additional piece of 2 cm × 2 cm N-Terface *on top* of each chimeric graft (at this point, the graft will be in between two pieces of 2 cm × 2 cm N-Terface).

(e) Tape the lid of each 150 mm dish to the bottom of the dish and place dishes into a resealable plastic bag.

(f) Place dishes in plastic bags into a Styrofoam box for transport to surgical facility.

(g) Upon arrival at surgical facility, store chimeric grafts in a 37 °C incubator until the time of grafting.

2. Preparation for grafting procedure:

This procedure is best performed by two surgeons (to maintain sterility) and one surgical assistant (not sterile).

(a) Surgical field.

- In a biological safety cabinet, set up two small heating pads. Next, use sterile surgical drapes to cover the surgical field, covering the heating pads and electrical wires.
- Add the following sterile items to the sterile surgical field:
  - Insulin syringe U-100 (Qty: number of animals to be grafted).
  - 1 mL TB syringe (Qty: number of animals to be grafted).
  - Tegaderm (Qty: number of animals to be grafted).
  - Steri-Strip Compound, Benzoin Tincture (Qty: number of animals to be grafted).
  - Suture, Ethilon® nylon, 6–0 (Qty: 2 suture packs per animal).
  - Coban™ self-adhering wrap (Qty: 2 roles).
  - 2 cm × 2 cm Cardboard square
  - Tongue depressor (Qty: 1).
  - Earhole punch (Qty: 1).
  - Needle driver (Qty: 1).
  - Sharp scissors (Qty: 2).
  - Blunt forceps (Qty: 2).
  - Fine forceps (Qty: 1).
  - Surgical marker (Qty: 1).
  - Standard 4″ × 4″ gauze (Qty: 3 packages).
  - Alcohol prep pads (Qty: 2 per animal).
  - Povidone-Iodine Prep Pads (Qty: 2 per animal).
  - Antibiotic ointment.

- While maintaining sterility, fill the insulin syringes completely with Avertin and fill the 1 mL syringes with saline.
- While maintaining sterility, stack four pieces of standard gauze together. Using a sharp scissors, cut the gauze into octagons approximately the size of the 2 cm × 2 cm chimeric graft. Generate enough octagons to correspond to the number of animals to be grafted.

(b) Procedure:

- Remove mouse from cage and administer 0.20–0.25 mL Avertin intraperitoneally (dependent on animal weight, estimated dose of 0.01 mL/g).
- Allow mouse to become fully anesthetized and confirm by toe pinch (typically requires 3–5 min).
- Position the anesthetized mouse to allow for grafting on the right lateral flank region. Prepare the surgical site by swabbing the dorsum of the mouse with 70% ethanol followed by povidone-iodine. Allow the area to dry.
- Using the surgical marker and 2 cm × 2 cm cardboard square, mark the corners and midpoints of the square to serve as a guide.
- Carefully snip the flank skin using sharp scissors following the guide marks down to the level of the panniculus carnosus. Apply a gentle forward, upward movement with the scissor tip, closing the scissors to further advance the incision. Use a fine pair of forceps to gently lift up the corner of the skin in the 2 cm × 2 cm region and gently scrape away the panniculus carnosus from the skin using the edge of the scissors, leaving the panniculus carnosus intact. Very little bleeding should occur. If there is voluminous bleeding, this indicates that the panniculus carnosus has been damaged which will compromise engraftment.
- Dispense ~2–4 mL of saline to irrigate the wound area.
- Remove 150 mm dish from incubator containing chimeric grafts and apply a graft covered with an overlying piece of non-adherent N-Terface to the full-thickness wound.
- To suture the graft in place, use a needle driver with the first pack of suture and begin by creating a stent-type suture at each of the four corners of the 2 cm × 2 cm square. Allow for ~8 cm to extend from each stitch.

- Using the second pack of suture and needle driver, create a stent-type stitch at each of the four midpoints of the 2 cm × 2 cm square. Allow for ~8 cm to extend from each stitch.
- Using a tongue depressor, apply antimicrobial ointment to one side of a gauze octagon and apply the octagon to the grafted area with antimicrobial ointment in contact with the wound. Ensure that 8 cm suture overhangs are lying outside of the grafted area to generate a stent stitch over the sterile gauze.
- Generate four total stent stitches by tying opposing sutures over the sterile gauze (i.e., corner to corner, midpoint to midpoint).
- Apply benzoin tincture along the wound perimeter. Quickly apply Tegaderm over the wound site, using the benzoin tincture to secure the Tegaderm in place.
- Administer 1 mL of saline as a resuscitation fluid intraperitoneally.
- Wrap the grafted wound using Coban, beginning cranially and working toward the caudal end. Use blunt forceps to "crimp" the free end of the Coban to ensure that it remains tightly fixed in place.
- Earhole punch the mouse (if required) and return to a heating pad to facilitate recovery from the anesthetic.
- Once the mouse becomes alert and active, return to housing cage.
- Repeat procedure as required.
- Maintain the mouse cages with half of the cage on a heating pad set on "low" for 2 weeks, and provide food and water *ad libitum*. Check mice daily for 2 weeks following the procedure to ensure that dressings remain intact. Rewrap mice with Coban as needed. Remove all forms of enrichment and ensure that food and water are readily and easily accessible.

3.5 Postsurgical Considerations

1. Chimeric follicular structures were histologically observable between 5 and 6 weeks after grafting. Hair shafts did not readily erupt to the skin surface; however, removal of the upper layer of epidermis will facilitate exposure of keratinized hair shafts near the skin surface (Fig. 2). Additional qualifications to the utility of this system for purposes of skin tissue engineering are provided below in section 4, Notes.

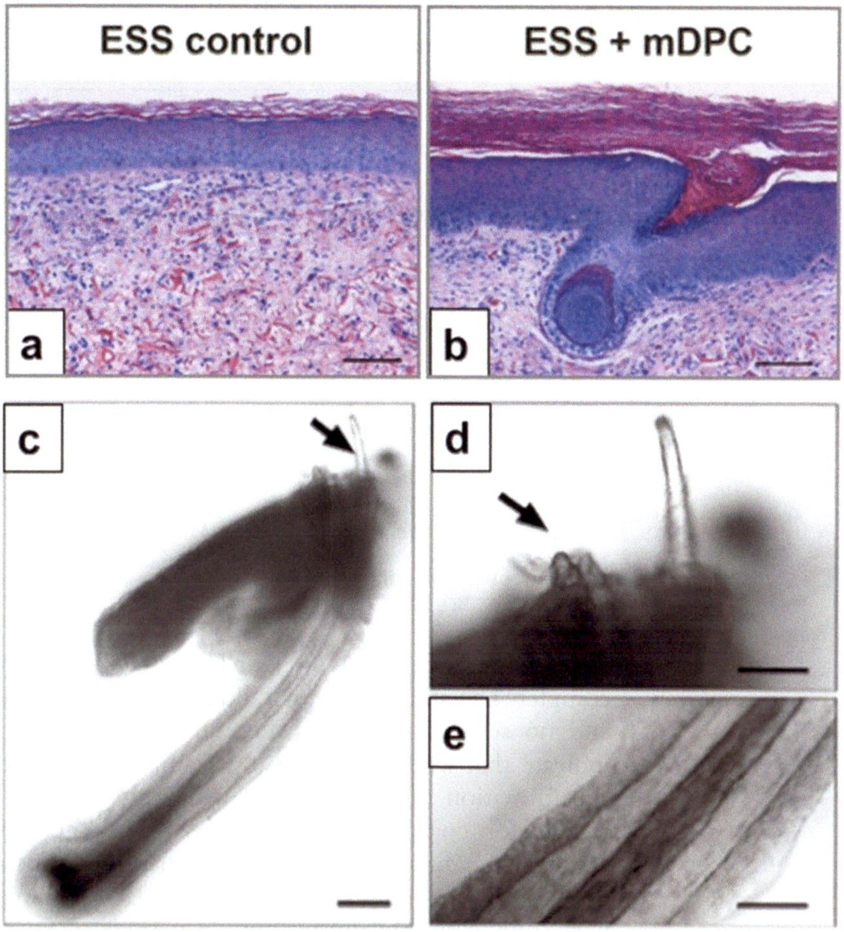

Fig. 2 Histological and morphological features of regenerated hair follicles in ESS with mDPC. Following grafting, follicular structures were observed histologically in ESS containing mDPC compared to control (**a**, **b**, scale bar = 100 μm). Gross inspection shows complete hair follicle with hair shaft eruption under skin surface (**c**, scale bar = 100 μm, **d–e**, scale bar = 50 μm). Images adapted from Ref. 9

4 Notes

1. This approach has been shown to produce follicular structures containing both murine and human cells; however, it is an incomplete pilosebaceous unit [9]. The generation of sebaceous glands, a key component of the pilosebaceous unit, is incomplete when utilizing this approach. The absence of sebaceous glands, and the subsequent lack of sebum production, presumably contributes to the inability of the hair shaft to erupt at the skin surface. Others have observed this as well. Higgins et al. created a model system in which hDPC spheroids were inoculated between foreskin epidermis and dermis and grafted

Table 2
Supplements for ESS maturation medium, UCMC160

	UCMC160, maturation medium [11]
Basal medium	DMEM
Epidermal growth factor	10 ng/mL
Insulin	5.0 μg/mL
Hydrocortisone	0.50 μg/mL
Progesterone	10 nM
Strontium chloride	1 mM
Linoleic acid	2.0 μg/nL
Triiodothyronine	20 pM
L-Ascorbic acid 2-phosphate	0.1 mM
Penicillin-streptomycin-fungizone	1%v/v

to an immunodeficient mouse. This system also produced a complete hair shaft although it did not erupt at the skin surface, and no sebaceous glands were observed [6]. These results indicate that while DPCs most certainly play a critical role in the formation (and cycling) of a hair follicle, the inductive signals for complete pilosebaceous morphogenesis do not reside in the differentiated DPC population. As mentioned in Subheading 1, the signals guiding follicular morphogenesis are likely only activated during limited periods of fetal morphogenesis, and are not usually expressed at postnatal time points. Furthermore, it has been shown that in vitro culturing of DPCs leads to decreased potency [6]. Future studies using this model system will seek to identify a target dermal cell population that remains potent to initiate morphogenesis of both pilosebaceous units and sweat glands into adulthood.

2. A key component of this model system is the use of optimized culture media and conditions to promote the growth, viability, and potency of the cell types used to generate the chimeric grafts. Preparation of these media types is extensive; however, the references provided describe detailed formulation protocols [11, 12]. *See* Tables 1 and 2 for media supplementations.

3. Isolation and transplantation of murine dermal papillae require a significant amount of practice and patience. Others have documented complete, detailed protocols which may be of use when performing this part of the procedure [16, 17]. The use of a high-resolution dissecting microscope is critical for success, as are steady hands.

In conclusion, controlled morphogenesis of chimeric hair follicles is an acquired skill that has significant value as a model of developmental biology with applications in tissue engineering and regenerative medicine. Knowledge gained from this and related models may be expected to contribute to elucidation of molecular pathways that regulate formation of epidermal appendages and normal physiology in uninjured skin. These findings will likely provide insights which can lead to reductions in morbidity and mortality in patient populations that suffer from acute and chronic wounds.

References

1. Holbrook KA, Minami SI (1991) Hair follicle embryogenesis in the human. Characterization of events in vivo and in vitro. Ann N Y Acad Sci 642:167–196
2. Muller M, Jasmin JR, Monteil RA, Loubiere R (1991) Embryology of the hair follicle. Early Hum Dev 26(3):159–166
3. Millar SE (2002) Molecular mechanisms regulating hair follicle development. J Invest Dermatol 118(2):216–225
4. Rendl M, Fuchs E (2005) Molecular dissection of mesenchymal-epithelial interactions in the hair follicle. PLoS Biol 3(11):1910–1924
5. Wu X, Scott L Jr, Washenik K, Stenn K (2014) Full-thickness skin with mature hair follicles generated from tissue culture expanded human cells. Tissue Eng Part A 20 (23–24):3314–3321
6. Higgins CA, Chen JC, Cerise JE, Jahoda CA, Christiano AM (2013) Microenvironmental reprogramming by three-dimensional culture enables dermal papilla cells to induce de novo human hair-follicle growth. Proc Natl Acad Sci U S A 110(49):19679–19688
7. Thangapazham RL, Klover P, Wang JA, Zheng Y, Devine A, Li S, Darling TN (2014) Dissociated human dermal papilla cells induce hair follicle neogenesis in grafted dermal-epidermal composites. J Invest Dermatol 134 (2):538–540
8. Sriwiriyanont P, Lynch KA, Maier EA, Hahn JM, Supp DM, Boyce ST (2012) Morphogenesis of chimeric hair follicles in engineered skin substitutes with human keratinocytes and murine dermal papilla cells. Exp Dermatol 21 (10):783–785
9. Sriwiriyanont P, Lynch KA, McFarland KL, Supp DM, Boyce ST (2013) Characterization of hair follicle development in engineered skin substitutes. PLoS One 8(6):e65664
10. Sriwiriyanont P, Maier EA, Lynch KA, Supp DM, Boyce ST (2011) Dermal papilla cells promote trichogenesis in engineered skin substitutes. J Investig Dermatol 131(S1):S79
11. Boyce ST, Rice RK, Lynch KA, Supp AP, Swope VB, Kagan RJ, Supp DM (2012) Assessment of replication rates of human keratinocytes in engineered skin substitutes grafted to athymic mice. Wound Repair Regen 20 (4):544–551
12. Boyce ST (1999) Methods for the serum-free culture of keratinocytes and transplantation of collagen-GAG-based skin substitutes. In: Morgan JR, Yarmush ML (eds) Methods in molecular medicine, Tissue engineering methods and protocols, vol 18. Humana Press Inc., Totowa, pp 365–389
13. Boyce S, Michel S, Reichert U, Shroot B, Schmidt R (1990) Reconstructed skin from cultured human keratinocytes and fibroblasts on a collagen-glycosaminoglycan biopolymer substrate. Skin Pharmacol 3:136–143
14. Boyce ST, Christianson DJ, Hansbrough JF (1988) Structure of a collagen-GAG dermal skin substitute optimized for cultured human epidermal keratinocytes. J Biomed Mater Res 22(10):939–957
15. Cunliffe-Beamer TL (1983) Biomethodology and surgical techniques. In: Foster HL, Small JD, Fox JG (eds) The mouse in biomedical research, vol III. Academic Press, New York, pp 417–418
16. Gledhill K, Gardner A, Jahoda CA (2013) Isolation and establishment of hair follicle dermal papilla cell cultures. Methods Mol Biol 989:285–292. https://doi.org/10.1007/978-1-62703-330-5_22
17. Jahoda C, Oliver RF (1981) The growth of vibrissa dermal papilla cells in vitro. Br J Dermatol 105(6):623–627

Chapter 14

Isolation and Culture of Epidermolysis Bullosa Cells and Organotypic Skin Models

Yinghong He and Cristina Has

Abstract

Isolation and culture of keratinocytes from patients with different types of epidermolysis bullosa are sometimes challenging, because of the inherent adhesion defects of these cells. We routinely employ a well-established protocol for in vitro culture of these cells from small skin samples remaining after diagnostic procedures are performed. Keratinocytes and fibroblast can be used for downstream expression and functional studies or for construction of in vitro organotypic cocultures. These cells maintain main common characteristics of spreading, adhesion, migration, and survival, which depend on the underlying molecular defect.

Key words Epidermolysis bullosa, Cell culture, Organotypic coculture, Adhesion, Blister, Keratinocyte, Fibroblast

Abbreviations

DMEM	Dulbecco's modified Eagle medium
EB	Epidermolysis bullosa
Fb	Fibroblasts
FCS	Fetal calf serum
PBS	Phosphate-buffered saline

1 Introduction

Isolation and culture of human skin cells from the same skin biopsy are the basic requisites for the research in the field of skin biology [1]. Human monogenetic diseases provide perfect paradigms for the investigation of the function of the single molecules in the skin cells and skin.

Epidermolysis bullosa (EB) is a heterogeneous group of genetic disorders characterized by mechanically induced skin blistering. EB

Sophie Böttcher-Haberzeth and Thomas Biedermann (eds.), *Skin Tissue Engineering: Methods and Protocols*, Methods in Molecular Biology, vol. 1993, https://doi.org/10.1007/978-1-4939-9473-1_14,

is caused by mutations in genes encoding structural proteins with roles in mechanical stability and cell-cell or cell-matrix adhesion, such as keratin5 and 14, type VII collagen, type XVII collagen, laminin 332, integrin α6β4, integrin α3, or kindlin-1 [2, 3]. All proteins which are altered in EB are expressed by keratinocytes, with exception of collagen VII, which is expressed by both keratinocytes and fibroblasts. Consequently, keratinocytes isolated from patients with EB display deficiencies in their adhesive function, the main property of epidermal keratinocytes. For example, laminin 332- and kindlin-1-negative cells hardly adhere on cell culture dishes, and are frequently lost during isolation from skin samples [4, 5]. To enable their culture, seeding on pre-coated dishes with extracellular matrix proteins (such as laminin 332 or Matrigel) is helpful. Intriguingly, keratinocytes isolated from patients with *COL7A1* or *COL17A1* mutations adhere quite well, but demonstrate other more subtle anomalies. Adhesion defects result in additional anomalies of cell survival and migration, which may influence cell culture. Because of these limitations in the culture of primary EB keratinocytes and their rapid senescence [6], we routinely immortalize primary cells by viral transduction of a plasmid containing the E6E7 HPV genes [7, 8]. Primary fibroblasts isolated from most EB patients behave like fibroblasts isolated from non-EB individuals, except of those from patients with dystrophic EB and *COL7A1* mutations and of those isolated from scars, which will display the corresponding features.

This chapter details the methods for extraction and culture of fibroblasts and keratinocytes from human skin biopsies, especially from EB patients, as used in our laboratory. They are followed by a detailed description of the generation of reconstructed skin by the self-assembly approach [9]. Other laboratories employ different methods or modifications of these protocols. The protocols included in this chapter are designed to enable rapid isolation of the cells from small biopsies in a daily routine. Skin samples, usually 4 mm punches, are obtained for molecular diagnostics by immunofluorescence mapping. After written informed consent from the patient, the material remaining after the diagnostic procedure is used for cell isolation for scientific purposes.

2 Materials and Methods

Prepare and store the culture media and the general materials in proper conditions and keep in sterile. Be sure to use analytical grade reagents for all the materials. Keep the disinfection rules and perform all procedures under the sterile hood. Diligently follow all waste disposal regulations when disposing waste materials and autoclave regulations.

2.1 Culture Media and Reagents

Sterile Dulbecco's modified Eagle medium (DMEM), 20 mM Hepes supplemented with 1% antibiotic/antimycotic mix.

2.1.1 Tissue Transport Medium

1. DMEM, 20 mM Hepes.
2. Antibiotic/antimycotic mix (100×). Distribute in aliquots and store at −20 °C until use.

Distribute in 5 mL aliquots and store at −20 °C until use.

2.1.2 Fb Medium (Human Fibroblast Culture Medium)

DMEM, 10% FCS, 1% antibiotic/antimycotic mix, 2% L-glutamine:

1. DMEM.
2. Antibiotic/antimycotic mix (100×).
3. Fetal calf serum (FCS).

Thaw in cold water. Distribute in aliquots and store at −20 °C until use.

4. 100 mL 200 mM L-Glutamine.

Thaw in cold water. Inactivate complement proteins in hot water (56 °C) for 30 min. Distribute in 10 mL aliquots and store at −20 °C until use.

5. Ascorbic acid: Dissolve 50 mg/mL ascorbic acid in sterile water to make a 1000× stock solution. Sterilize by filtration through a 0.22 μm low-binding disposable filter (*see* **Note 1**).

2.1.3 KGM Medium (Human Keratinocyte Culture Medium)

Keratinocyte SFM with supplements:

1. Keratinocyte serum-free medium: 500 mL SFM.
2. Supplements: 2.5 μg Human recombinant EGF, 25 mg bovine pituitary extract.

2.1.4 OTC: Medium (Organotypic Coculture Medium)

Air-liquid medium, supplements:

1. Air-liquid medium: 3:1 DMEM/Ham's F12, 5% fetal bovine serum, 100 U penicillin, 100 μg/mL streptomycin.
2. Supplements: 5 μg/mL Insulin, 1×10^{-4} M adenine, 1×10^{-10} M cholera toxin, 0.4 μg/mL hydrocortisone, 1 ng/mL EGF.

2.1.5 Freezing Medium

Cell growth medium, 20% FCS, 10% DMSO:

1. Fetal calf serum.
2. Dimethyl sulfoxide (DMSO).

2.1.6 General Materials

1. Phosphate-buffered saline (PBS).
2. Trypsin/EDTA solution (0.05/0.02% w/v), 100 mL.

Thaw in cold water. Distribute in 10 mL aliquots and store at −20 °C until use.

3. Falcon tubes, Petri dish, and parafilm.
4. Dissecting curved forceps.
5. Scalpel and blade.
6. Tissue culture flask, 75 cm^2.
7. Sterile cryogenic vials.
8. Deep-well plates.
9. Filter inserts, pore size 1 μm.
10. Glass rings (1, 2 cm diameter, 1 cm high).
11. Sterile glass beaker with magnetic stirrer.
12. Hanks' balanced salt solution (HBSS).
13. Collagen type I (acid extracted from rat tail, concentration 6 mg/mL).

2.2 Monolayer Culture of the Fibroblasts from the Skin Biopsy from the EB Patient

2.2.1 Extraction and Culture of Human Fibroblasts

1. Source of cells: A small piece of skin biopsy from EB patient (fresh blister or after rubbing the skin) removed by surgery following procedures approved by the institution's committee for the protection of human subjects can be stored in the tissue transport medium at 4 °C for a maximum of 3 days.
2. Remove excess fat and blood, and rinse tissue many times with PBS supplemented with 8% antibiotic/antimycotic mix.
3. Cut skin into pieces of 2 × 2 mm size.
4. Transfer pieces into empty T25 cell culture flask and place dermis side down.
5. Incubate for 30–60 min at 37 °C in a cell culture incubator to allow tissue to adhere to the bottom of the flask.
6. After tissue is well attached, add 5 mL in Fb medium, and incubate at 37 °C, 5% CO_2.
7. First medium change is after 5 days; usually fibroblasts grow out of the tissue within 7 to 14 days, and sometimes it can take up to 30 days (Fig. 1).

2.2.2 Subculture of Human Fibroblasts (Passage)

1. Discard the medium, and wash cells with PBS.
2. Add 1–2 mL trypsin solution (2.5% w/v, diluted 1:10 with PBS) per T25 cell culture flask and 3 mL per T75 cell culture flask.
3. Cells need about 5 min to detach; watch continuously to avoid overlong incubation.
4. Stop reaction by adding 2–3 times the volume of PBS with 10% FCS.

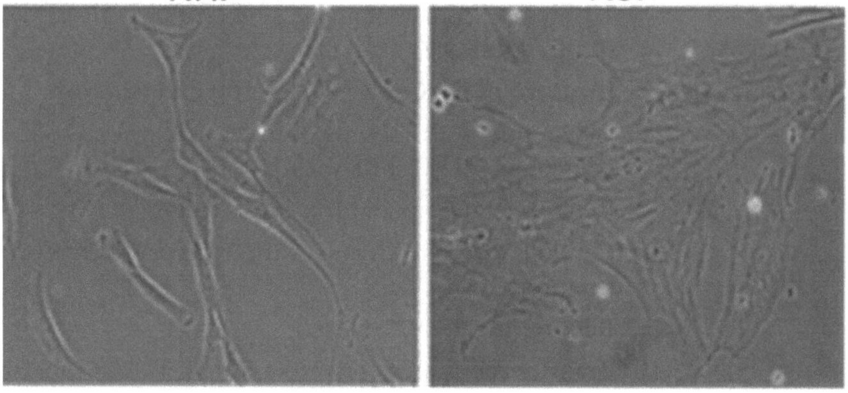

Fig. 1 Morphology of primary normal human (NHF, left) and Kindler syndrome (KSF, right) fibroblasts. Note that Kindler syndrome fibroblasts are larger and display morphology suggestive for myofibroblasts

5. Transfer cell suspension into a Falcon tube, and centrifuge for 5 min at 1000 × *g*, at room temperature (RT).
6. Resuspend the pellet in Fb medium.
7. Plate approximately 0.5×10^6 cells per T25 cell culture flask in Fb medium.

2.2.3 Cryopreservation of Human Fibroblasts

1. Resuspend 1×10^6 cells in 1 mL freezing medium.
2. Freeze at −80 °C in an isopropanol-filled freezing container.
3. Transfer to liquid nitrogen within 48 h after freezing.

2.2.4 Thawing of Human Fibroblasts

1. Thaw quickly the 1 mL cryogenic vial containing the milliliter cell-liquid from liquid nitrogen storage in a 37 °C water bath until the milliliter is liquid.
2. Resuspend in 15 mL cell culture medium, wash cells, and centrifuge the cells in 1000 × *g* for 5 min.
3. Collect the cell pellet in the appropriate volume of culture medium and plate on a cell culture flask.

2.3 Monolayer Culture of the Keratinocytes from Skin Biopsies from EB Patients

2.3.1 Extraction and Culture of Human Keratinocytes

1. Source of cells: A small piece of skin biopsy from EB patient removed by surgery following procedures approved by the institution's committee for the protection of human subjects can be stored in the tissue transport medium at 4 °C for a maximum of 3 days.
2. Remove excess fat and blood, and rinse tissue many times with PBS supplemented with 8% antibiotic/antimycotic mix.
3. Transfer skin sample to sterile Petri dish and remove dermis as well as possible.

4. Cut skin sample into 5 × 5 mm pieces and transfer them to a 10 cm Petri dish containing 30 mL of trypsin/EDTA solution (0.05/0.02% w/v).
5. Incubate at 37 °C, 5% CO_2, for 60 min (*see* **Note 2**).
6. When epidermis loosens, transfer skin pieces to a Petri dish containing 30 mL PBS with 10% FCS.
7. Remove epidermis and tear into small fragments with forceps.
8. Carefully scratch dermis with forceps to remove residual keratinocytes.
9. Remove the dermal tissue.
10. Transfer cell suspension into a 50 mL Falcon tube and centrifuge for 5 min at 1000 × *g*, RT.
11. Resuspend pellet in KGM medium (*see* **Note 3**) and plate into T25 cell culture flask or in one 6-well (approximately 1 cm^2 skin/T25) (*see* **Note 4**).
12. Incubate at 37 °C, 5% CO_2.
13. Observe cells daily and handle cells gently, especially keratinocytes deficient for proteins of the dermal-epidermal junction (Fig. 2).
14. Change medium three times a week.

2.3.2 Subculture of Human Keratinocytes (Passage)

1. Discard the medium.
2. Add 1–2 mL trypsin solution (2.5% w/v, diluted 1:10 with PBS) per T25 cell culture flask and 3 mL per T75 cell culture flask.
3. Cells need about 5 min to detach; watch continuously to avoid overlong incubation.
4. Stop reaction by adding 2–3 times the volume of PBS with 10% FCS.

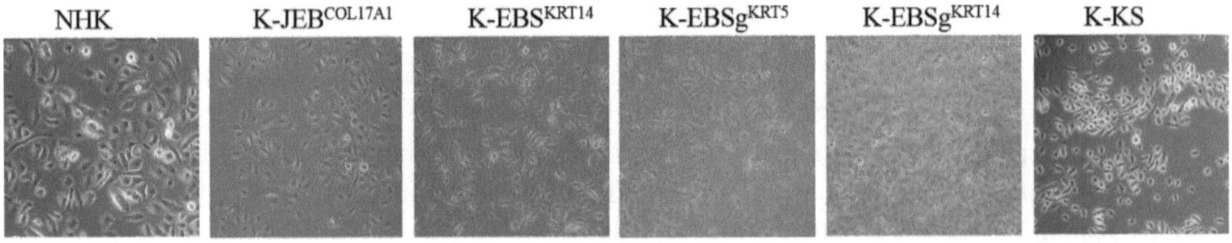

Fig. 2 Morphology of primary normal human (NHK) and EB keratinocytes: junctional EB with *COL17A1* mutations, K-JEB^{COL17A1}, EB simplex localized with *KRT14* mutation K-EBSl^{KRT14}, EB simplex generalized with *KRT5* mutation K-EBSg^{KRT5}, EB simplex generalized with *KRT14* mutation K-EBSg^{KRT14}, Kindler syndrome keratinocyte K-KS

5. Transfer cell suspension into a Falcon tube, and centrifuge for 5 min at 1000 × *g*, RT.
6. Resuspend the pellet in KGM medium and count the cells.
7. Plate approximately 0.5 × 10^6 cells per T25 in KGM medium.

2.3.3 Cryopreservation of Human Keratinocytes

As described in Subheading 2.2.3.

2.3.4 Thawing of Human Keratinocytes

As described in Subheading 2.2.4.

2.4 Human Skin Reconstruction by the Self-Assembly Approach

All further manipulations are performed under a sterile laminar flow hood cabinet.

2.4.1 Formation of Fibroblast-Collagen I Gel for the Dermal Sheets

1. Put the filter inserts in each well into the 6-well deep-well culture plate (*see* **Note 5**).
2. Mix collagen I solution with HBSS, and then neutralize the mixture by adding 5 M NaOH.
3. Confluent fibroblasts between their second and sixth passages will be suspended in the FCS in a concentration with 3 × 10^6/mL gel.
4. Add FCS-fibroblast suspension to collagen solution; mix thoroughly in the sterile glass beaker with magnetic stirrer.
5. Pipet 2.5 ml of the mixture into one filter insert (*see* **Note 6**) and incubate at 37 °C, 5% CO_2.
6. Gels solidify within 1 h at 37 °C, set glass ring into insert, and press down gently without rupturing the gel.
7. After 1 h remove the fluid and immerse gel in Fb medium containing 50 μg/mL ascorbic acid (Fig. 3a).

2.4.2 Human Skin Reconstruction (Keratinocyte Addition)

1. One day after the preparation of the gels, remove medium inside the glass ring.
2. Resuspend the trypsinized EB keratinocytes as described in KGM medium in 1 × 10^6/mL.
3. Remove the culture medium from reconstructed dermis. Seed 1 × 10^6 keratinocytes (1 mL) into the glass ring. Incubate at 37 °C, 5% CO_2.
4. One day later after seeding of keratinocytes, the glass ring can be gently removed with forceps (*see* **Note 7**).

2.4.3 Maturation of the Skin Model: Air-Liquid Interface Culture

1. One day after the seeding of keratinocytes, remove culture medium above the insert.
2. Lift reconstructed skin and transfer it on the air-liquid stand.

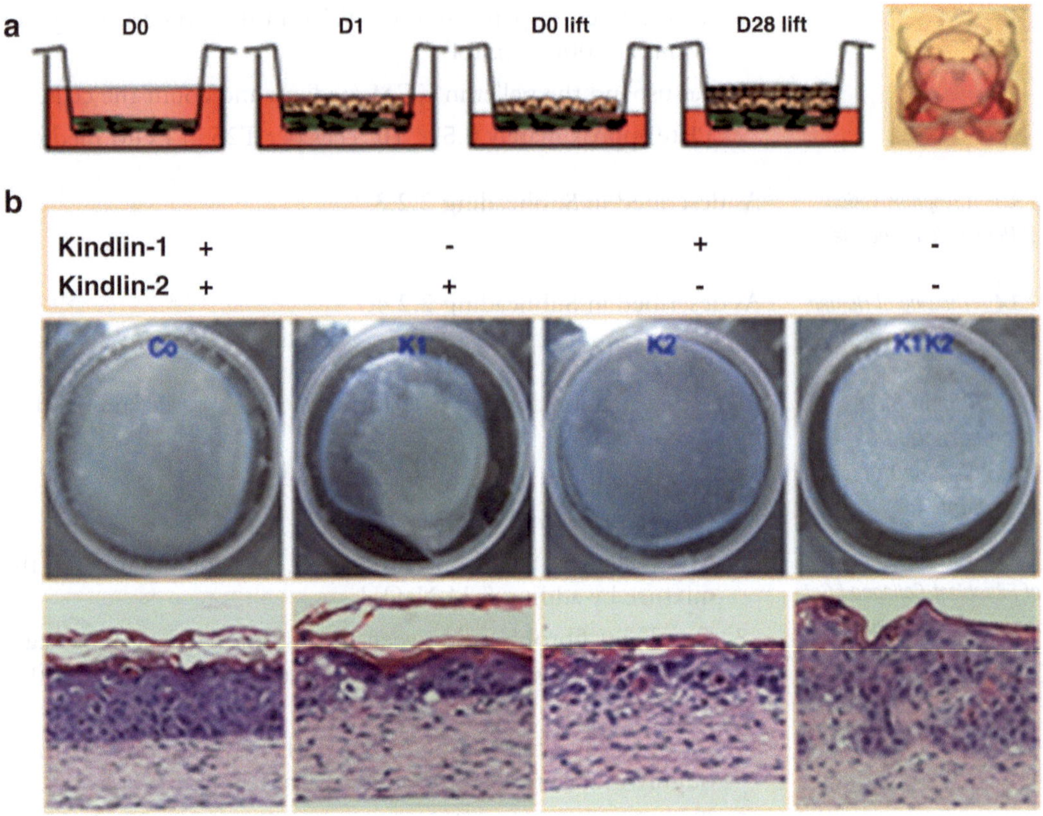

Fig. 3 (**a**) Schematic representation of the method for in vitro skin reconstruction; green, dermal equivalent, yellow epidermal equivalent; the right panel shows a picture of an organotypic skin coculture. (**b**) Reconstructed skin with normal fibroblasts and kindlin-1, or kindlin-2, or kindlin-1- and -2-deficient keratinocytes. Upper panels: reconstructed skin in culture; lower: histological section of human reconstructed skin after 2 weeks. Note anomalies in the epidermal structure when the kindlin-deficient keratinocytes were used

3. Add 10 mL of OTC medium containing 50 μg/mL ascorbic acid (*see* **Note 8**). Incubate at 37 °C, 5% CO_2. Change culture medium three times a week.
4. After 2 weeks, reconstructed skin is then ready for various types of analysis after removal from the insert filters.
5. The reconstructed skin tissues will be snap frozen or formalin fixed for the further analysis. According to the need of the experiment, the reconstructed skin can be cultivated at the air-liquid interface for more than 28 days. However, a well-organized basement membrane is already obtained after 14 days of culture at the air-liquid interface (Fig. 3b).

3 Notes

1. Ascorbic acid stock solution must be prepared immediately before use and protected from light.
2. The incubation times are depending on tissue type; for the blister roof we usually use 20–35 min, for a full-thickness skin at 60 min, and for the foreskin at 60–90 min.
3. Preincubate KGM medium in the cell culture incubator for about 1 h; the keratinocytes will attach and spread more efficiently.
4. Coating the cell culture flasks with different feeder layer sometimes improves the attachment and spreading of the EB keratinocytes.
5. For the skin model construction usually will be performed in 6-well plates. However, they have already performed this model using 12-well plates according to the need of the experiment. The size of the cell culture dish to be used is thus not limited and is to be chosen according to the desired type of reconstruction and experiment to be performed. Therefore, filter inserts and deep-well culture plates and glass rings should be adjusted for that particular type of cell culture dish.
6. Take care not to produce any air bubbles; preincubating the pipet used for this step in minus 20° helps for the manipulation.
7. Place one end of the plate higher so that you can see the epithelium within the ring. Gently turn the glass ring using two sterile forceps to remove glass ring, and take care not to disturb the epithelium.
8. The lower surface of the reconstructed skin must be in direct contact with culture medium.

Acknowledgments

The authors would like to thank current and former members of the laboratory. A special thanks is said to Juna Leppert team members who have contributed to adapt the protocols to obtain this for her valuable suggestions in the manuscript.

References

1. Germain L, Rouabhia M, Guignard R, Carrier L, Bouvard V, Auger FA (1993) Improvement of human keratinocyte isolation and culture using thermolysin. Burns 19:99–104
2. Bruckner-Tuderman L, Has C (2014) Disorders of the cutaneous basement membrane zone--the paradigm of epidermolysis bullosa. Matrix Biol 33:29–34

3. Has C, He Y (2017) Focal adhesions in the skin: lessons learned from skin fragility disorders. Eur J Dermatol S1:8–11
4. Gagnoux-Palacios L, Vailly J, Durand-Clement-M, Wagner E, Ortonne JP, Meneguzzi G (1996) Functional Re-expression of laminin-5 in laminin-gamma2-deficient human keratinocytes modifies cell morphology, motility, and adhesion. J Biol Chem 271:18437–18444
5. Herz C, Aumailley M, Schulte C, Schlötzer-Schrehardt U, Bruckner-Tuderman L, Has C (2006) Kindlin-1 is a phosphoprotein involved in regulation of polarity, proliferation, and motility of epidermal keratinocytes. J Biol Chem 281:36082–36090
6. Piccinni E, Di Zenzo G, Maurelli R, Dellambra E, Teson M, Has C, Zambruno G, Castiglia D (2013) Induction of senescence pathways in Kindler syndrome primary keratinocytes. Br J Dermatol 168:1019–1026
7. He Y, Maier K, Leppert J, Hausser I, Schwieger-Briel A, Weibel L, Theiler M, Kiritsi D, Busch H, Boerries M et al (2016) Monoallelic mutations in the translation initiation codon of KLHL24 cause skin fragility. Am J Hum Genet 99:1395–1404
8. Maier K, He Y, Esser PR, Thriene K, Sarca D, Kohlhase J, Dengjel J, Martin L, Has C (2016) Single amino acid deletion in kindlin-1 results in partial protein degradation which can be rescued by chaperone treatment. J Invest Dermatol 136:920–929
9. Stark H-J, Szabowski A, Fusenig NE, Maas-Szabowski N (2004) Organotypic cocultures as skin equivalents: a complex and sophisticated in vitro system. Biol Proced Online 6:55–60

Part III

In Vitro and Vivo Models Testing Skin, Skin Cells, and Tissue Engineered Skin

Chapter 15

Effects of the Extracellular Matrix on the Proteome of Primary Skin Fibroblasts

Regine C. Tölle and Jörn Dengjel

Abstract

The cellular microenvironment often plays a crucial role in disease development and progression. In recessive dystrophic epidermolysis bullosa (RDEB), biallelic mutations of the gene *COL7A1*, encoding for collagen VII, the main component of anchoring fibrils, lead to a loss of collagen VII in the extracellular matrix (ECM). Loss of collagen VII in skin is linked to a destabilization of the dermal-epidermal junction zone, blister formation, chronic wounds, fibrosis, and aggressive skin cancer. Thus, RDEB cells can serve as a model system to study the effects of a perturbed ECM on the cellular proteome. In this chapter, we describe in detail the combination of stable isotope labeling by amino acids in cell culture (SILAC) of primary skin fibroblasts with reseeding of fibroblasts on decellularized collagen VII-positive and -negative ECM to study the consequences of collagen VII loss on the cellular proteome. This approach allows the quantitative, time-resolved analysis of cellular protein dynamics in response to ECM perturbation by liquid chromatography-mass spectrometry.

Key words Skin, Fibroblasts, Protein kinetics, Decellularization, Extracellular matrix, Proteomics, GASP, High-pH reversed-phase chromatography, SILAC, Mass spectrometry

1 Introduction

The extracellular matrix (ECM) is of critical importance for cell and tissue homeostasis and its dysregulation is linked to the progression of numerous diseases. In accordance, the influence of the ECM on intracellular signaling has increasingly become the focus of research [1]. Creating an acellular matrix scaffold is a widely used method to study the physiological role and function of the microenvironment on cells, in tissues, and organs, and its contribution to diseases [2, 3]. Decellularized ECM can be repopulated by cells promoting cell proliferation and differentiation [4]. Importantly, it can be employed to study the kinetics of cell responses addressing how cells perceive signals, and how signal initiation and transduction are orchestrated leading to altered gene expression, and resulting in qualitative and/or quantitative changes of the proteome.

Sophie Böttcher-Haberzeth and Thomas Biedermann (eds.), *Skin Tissue Engineering: Methods and Protocols*, Methods in Molecular Biology, vol. 1993, https://doi.org/10.1007/978-1-4939-9473-1_15,

Recessive dystrophic epidermolysis bullosa (RDEB) is an inherited skin fragility disorder caused by biallelic mutations in the gene *COL7A1* resulting in the loss of collagen VII [5]. Collagen VII forms anchoring fibrils which support dermal-epidermal adhesion [6]. Its loss is linked to a complex disease phenotype. Respective individuals suffer from trauma-induced skin blistering, causing subsequent scarring, chronic wounds, and aggressive skin cancer [7, 8]. By mass spectrometry (MS)-based proteomics it was shown that RDEB fibroblasts produce a globally altered cellular microenvironment [9], which appears to actively contribute to the progression of aggressive squamous cell carcinoma [7, 9, 10]. However, loss of collagen VII seems to alter not only the cellular microenvironment but also the intracellular proteome [11, 12]. Whether this is due to a perturbed ECM or due to intracellular alterations linked to collagen synthesis and secretion remains to be studied.

Since the molecular consequences linked to the loss of collagen VII in the ECM are known [9], RDEB cells are a suitable model to study how perturbation of the ECM influences cellular signaling and decision finding. Here we describe a strategy to quantitatively analyze intracellular proteome alterations in response to interactions of cells with decellularized matrices isolated from control or RDEB fibroblast. For quantitation by MS, we implement stable isotope labeling by amino acids in cell culture (SILAC) for primary human skin fibroblasts. Additionally, we present a protocol for the fractionation of complex proteome samples in large volumes prior to MS analysis to ensure a deep proteome coverage.

2 Materials

2.1 Cell Culture

1. Dulbecco's modified Eagle medium (DMEM), high glucose 4.5 g/L.
2. 10% Fetal bovine serum (FBS).
3. SILAC-DMEM (high glucose 4.5 g/L), without L-lysine and L-arginine.
4. 10% Dialyzed fetal bovine serum (dFBS).
5. 200 mM L-Glutamine (100× stock solution).
6. 10,000 U/mL Penicillin, 10 mg/mL streptomycin (100× stock solution).
7. Following SILAC labels are used: L-lysine-$^{2}H_4$, L-arginine-$^{13}C_6$-$^{14}N_4$ (Lys4, Arg6), L-lysine-$^{13}C_6$-$^{15}N_2$, and L-arginine–$^{13}C_6$–$^{15}N_4$ (Lys8, Arg10) (*see* **Note 1**). For primary fibroblasts, add 84 mg/L L-arginine, 146 mg/L L-lysine (all from EURISO-TOP GmbH), and 164 mg/L proline (Sigma-Aldrich, *see* **Note 2**).
8. Ascorbic acid.

9. Trypsin-EDTA solution (200 mg/L trypsin, 500 mg/L EDTA).
10. Tissue culture flasks.
11. Sterile phosphate-buffered saline (PBS).
12. Syringe-driven filters, 0.2 μm, 30 mm diameter.
13. Primary normal human fibroblasts (NHF) from foreskin.
14. Primary fibroblasts from RDEB patients.

2.2 ECM Decellularization and Cell Harvest

1. 0.5% Triton X-100 in 20 mM NH_4OH.
2. 4% SDS, 0.1 M Tris–HCl (pH 7.6), 1 μM DTT.

2.3 Gel-Aided Sample Preparation (GASP)

1. 40% Acrylamide/Bis-acrylamide, tetramethylethylenediamine (TEMED), ammonium persulfate (APS).
2. 50% Methanol, 10% acetic acid, 40% dH_2O.
3. 6 M Urea in Tris–HCl, adjusted to pH 7.6.
4. Acetonitrile, MS grade.
5. ABC buffer: 100 mM Ammonium bicarbonate, pH 7.5.
6. 5 μg Lysyl endopeptidase (Waco) in 100 mM ABC buffer per gel plug.
7. 40 μg Sequencing-grade modified trypsin (Promega) per gel plug.
8. 5% Formic acid (MS grade) in dH_2O.

2.4 High-pH Reversed-Phase Chromatography

1. RP buffer A, pH 10: 10 mM Ammonium formate in dH_2O, adjusted to pH 10 with ammonia.
2. ReproSil-Pure C18-basic (Dr. Maisch) in methanol (MS grade).
3. C18 disks (3M Empore).
4. 200 μL Pipet tips.
5. Acetonitrile in RP buffer A, pH 10: 2%, 6%, 10%, 12%, 13%, 14%, 16%, 20%, 25%, and 50% acetonitrile for step elution of peptides.
6. Buffer A: 0.5% Acetic acid (MS grade) in dH_2O.
7. Buffer A*: 0.3% Trifluoroacetic acid in 3% acetonitrile (all MS grade) in dH_2O.
8. Buffer B: 80% Acetonitrile, 0.5% acetic acid in dH_2O.

3 Methods

In this protocol, we generate control and RDEB ECM in vitro and use it as a scaffold for NHF and RDEB fibroblasts (Fig. 1a). For quantitation of protein dynamics in cells seeded on decellularized ECM, we implemented SILAC-based MS. MS samples are prepared by GASP with subsequent high-pH reversed-phase chromatography for peptide fractionation (Fig. 1b) [13–15]. In our experimental setup, we mixed heavy-labeled NHF, which were cultured on RDEB ECM, with the medium-labeled NHF cultured on control ECM. Complementarily, we mixed heavy-labeled RDEB cells, which were cultured on control ECM, with medium-labeled RDEB cells cultured on RDEB ECM. Heavy/medium SILAC ratios imply cells cultured on their own ECM in the denominator. With this approach, protein abundances are normalized to the cellular response to reseeding and significant outliers should solely reveal the effects of RDEB ECM on NHF and of control ECM on RDEB cells, respectively (Fig. 1a). The experimental design can be changed, depending on the questions asked: for example NHF cultured on control ECM can be mixed with RDEB cells seeded on control ECM to study the different effects the same ECM has on distinct cell types in single-MS experiments.

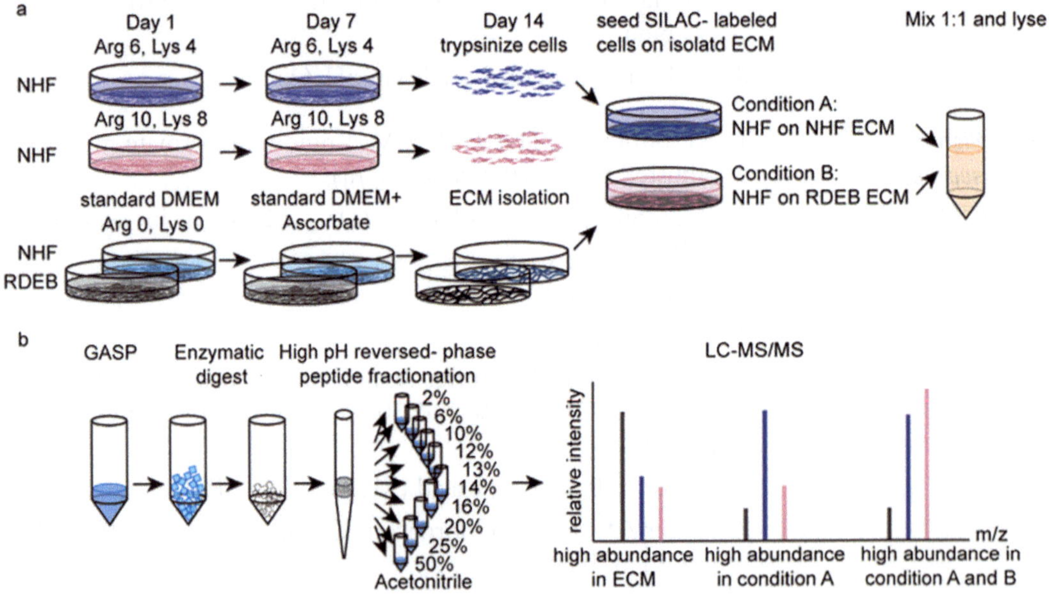

Fig. 1 Schematic representation of the experimental procedure for a SILAC-based ECM reseeding experiment followed by LC-MS/MS analysis. (**a**) Cells are SILAC labeled and seeded onto fibroblast-generated ECM. After a defined incubation period, the different conditions are mixed and samples are prepared for LC-MS/MS analysis. (**b**) Proteins are digested by gel-aided sample preparation (GASP) using trypsin and lysyl endopeptidase. High-pH reversed-phase chromatography with ten fractions is utilized for sample fractionation prior to LC-MS/MS analysis

3.1 SILAC Labeling of Primary Skin Fibroblasts

1. To transfer cells from standard DMEM to SILAC DMEM, wash cells with PBS, trypsinize, and spin down for 3 min at 300 × *g*, RT. Discard the supernatant.
2. Take up fibroblasts in SILAC-DMEM, supplemented with 10% dFBS, 2 mM L-glutamine, 100 U/mL penicillin, 0.1 mg/mL streptomycin, 84 mg/L L-arginine, 146 mg/L L-lysine, and 164 mg/L proline (*see* **Notes 2** and **3**).
3. For sufficient labeling, cells should be cultured in SILAC-DMEM for at least seven cell doublings [16]. Cells should not reach a confluence of 100% to ensure an active cellular metabolism and incorporation of the isotopically labeled amino acids.
4. Change media every other day.
5. Check labeling efficiency and proline conversion before starting large-scale experiments.

3.2 ECM Generation and Isolation

1. For ECM generation, culture cells in standard DMEM with high glucose, 10% FBS, 2 mM L-glutamine, 100 U/mL penicillin, and 0.1 mg/mL streptomycin.
2. Use 10 cm plates with 5 × 10^5–1 × 10^6 cells per plate, depending on the cell type.
3. For proper collagen production, treat cells for 7 days in total with 50 μg/mL ascorbic acid [9, 16] (*see* **Note 4**). Change medium every other day with freshly added ascorbic acid.
4. For ECM isolation work on ice: wash the cells three times with PBS.
5. Remove cells from ECM by washing with 0.5% Triton X-100 and 20 mM NH_4OH in dH_2O [17]. Add 1 mL per 10 cm plate for approximately 30 s. Check the efficiency of cell removal under the microscope.
6. Add 10 mL PBS and aspirate the solution.
7. Wash three times with PBS and add again 1 mL of 0.5% Triton X-100 and 20 mM NH_4OH in dH_2O.
8. Immediately add PBS and aspirate the solution.
9. Wash again five times with PBS to remove remaining intracellular debris, detergent, and ammonium hydroxide. Add DMEM to plates and store them for further use.

3.3 Reseeding of Fibroblasts on ECM

1. Trypsinize the fully labeled SILAC fibroblasts for 10 min at 37 °C, 5% CO_2.
2. Carefully aspirate the medium from the ECM.
3. Seed 5 × 10^5–1 × 10^6 cells in 10 mL SILAC-DMEM per 10 cm plate on top of the ECM (*see* **Note 5**). As example, seed NHF of one SILAC label on control ECM and NHF of

another SILAC label on RDEB ECM. Do the same with RDEB cells.

4. Keep cells and ECM in SILAC media at 37 °C, 5% CO_2, until lysis.

3.4 Gel-Aided Sample Preparation

1. Harvest samples with 0.5 mL 4% SDS, 0.1 M Tris–HCl (pH 7.6), and 1 μM DTT.
2. Heat and shake lysates for 5 min at 95 °C.
3. Mix opposing SILAC labels with each other in a 1:1 ratio. Each vial should contain three SILAC labels from two conditions, i.e., Arg0, Lys0: ECM; Arg6, Lys4: NHF on control ECM; and Arg10, Lys8: NHF on RDEB ECM (Fig. 1).
4. Add acrylamide/Bis-acrylamide to the lysate up to a final concentration of 20%. Mix sample by pipetting and incubate for 20 min at RT.
5. Add tetramethylethylenediamine (TEMED) to a concentration of 5% and ammonium persulfate to a final concentration of 0.5% w/v. Let the sample rest until it solidifies.
6. Shred the gel plug (*see* **Note 6**).
7. Fix gel pieces with 50% methanol, 10% acetic acid, and 40% dH_2O for 10 min. Samples should be entirely covered by the fixing solution. Shake carefully while fixing.
8. Wash once with 6 M urea and once with acetonitrile to remove detergents.
9. Wash twice, alternatingly with 100 mM ABC buffer and acetonitrile.
10. This step is optional to increase identification rates: incubate dehydrated gel pieces with 5 μg lysyl endopeptidase in 100 mM ABC buffer for 1 h at 37 °C [18].
11. Add 40 μg trypsin per gel plug and incubate samples overnight at 37 °C.
12. Collect all the peptides by extracting them from the gel: add 1 mL of 5% formic acid to stop the tryptic reaction. Dehydrate gel pieces twice with 1 mL acetonitrile. Collect all the supernatants in a single reaction tube.
13. Reduce the samples to less than 300 μL in a vacuum concentrator and add 1.5 mL RP buffer A pH 10.

3.5 High-pH Reversed-Phase Chromatography

1. Reversed-phase columns are self-packed: stack four layers of C18 disks in a 200 μL pipet tip. Add slurry of ReproSil-pure C18-basic in methanol on top to form a layer of approx. 3 mm C18 material (*see* **Note** 7). The pipet tip column should not get dry.

2. Wash column twice with 100 μL buffer B by centrifuging tips in reaction tubes in a tabletop centrifuge (~2 min, 3000 × *g*).
3. Equilibrate twice with 100 μL RP buffer A, pH 10, before the sample is loaded onto the column.
4. Load the sample onto the column by centrifugation.
5. Wash the column with 100 μL RP buffer A, pH 10.
6. Prepare ten tubes with 3 μL of 5% formic acid.
7. Add 50 μL of RP buffer A, pH 10, and 2% acetonitrile to the column. Centrifuge the tip for 1.5–2 min at 3000 × *g* and collect the flow through in a tube with 5% formic acid. Formic acid acidifies and thus stabilizes the peptides and potential posttranslational modifications.
8. Continue adding 50 μL RP buffer A, pH 10, with 6%, 10%, 12%, 13%, 14%, 16%, 20%, 25%, and 50% acetonitrile. Collect the respective flow through in separate tubes containing 5% formic acid.
9. Evaporate solvents in a vacuum concentrator to remove ammonium formate and acetonitrile.
10. Suspend the peptides in 15 μL of 30% buffer A* and 70% buffer A and store at −80 °C for LC-MS/MS analysis [19].

3.6 Data Analysis

1. Decellularization of fragile ECM is technically challenging and has to be optimized. Western blot analysis of different cellular components to analyze the purification efficacy is advisable (Fig. 2a). We used Histone H2B, Tenascin-C, and GAPDH as readout for nuclear, extracellular, and cytoplasmic cell fractions, respectively. Decellularized ECM should contain Tenascin-C and be free of GAPDH or Histone H2B.
2. Bioinformatics interpretation of MS data and statistical analyses can be done by the freely available software Perseus [20].

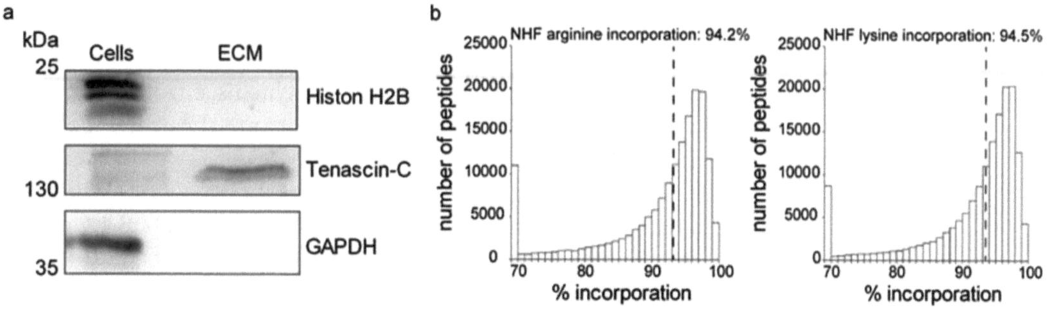

Fig. 2 Quality assessment of decellularized ECM and labeling efficiency. (**a**) Western blot analysis of trypsinized cells and isolated ECM. (**b**) Incorporation efficacy of isotopically labeled arginine and lysine of NHF, which were cultured on decellularized, non-labeled ECM. Dotted lines represent median label incorporation

3. The SILAC labeling efficiency should be determined to ensure complete labeling and accurate MS-based quantification (*see* **Note 8**). All non-labeled peptides should be associated with the decellularized ECM and are excluded from the subsequent data analysis as follows. All peptides are annotated according to the matrisome classification [21]. Peptides associated with "matrisome protein," "basal membrane protein," and "collagen" are removed for the determination of labeling efficiency. For the remaining peptides, the ratio of the intensity of the labeled peptide compared to the total intensity is calculated (Fig. 2b). Incorporation rates of 94% are sufficient since technical MS-quantitation errors of 10–20% are common.
4. Here we cultured cells on control and decellularized RDEB ECM for 30 min, 6 h, 12 h, and 24 h. All time points are normalized relative to the 30-min time point. Ratios are analyzed by a two-sided t-test with a Benjamini-Hochberg-corrected false discovery rate of 0.05. Only proteins with a significant ratio in at least one time point are submitted to z-normalization and k-means clustering. K-means clustering is performed with the freely available "Multi Experiment Viewer, MeV" [22].
5. GO term enrichment of the different clusters can be done by STRING v10.5 [23]. Depicted in Fig. 3 are 6 out of 15 clusters. Proteins in clusters 1 and 11 are upregulated in RDEB fibroblasts in response to NHF ECM but show a negative or no response in NHF cultured on RDEB ECM. These clusters mainly contain proteins regulating ECM-receptor interaction, proteoglycans in cancer, focal adhesion, amoebiasis, and PI3K-Akt signaling pathway. Cluster 1 contains DCN, TGFBI, COL6A2, and SERPINE2, which are known to be upregulated in RDEB fibroblasts [9], indicating that the increased TGFβ signaling observed in RDEB [24, 25] is cell autonomous and might depend on intracellular sensing of collagen VII. Also, the basement membrane-associated proteins COL4A2 and LAMB1 are increasingly expressed in RDEB fibroblasts in response to NHF ECM. COL4A2 and LAMB1 are known to be downregulated in RDEB ECM [9], which seems to be rescued by culturing RDEB fibroblasts on healthy collagen VII-positive ECM.

   On the contrary, clusters 5 and 10 show proteins which are increasingly expressed in NHF in response to RDEB ECM but negatively or do not respond in RDEB fibroblasts cultured on control NHF ECM. These proteins are mainly involved in DNA replication, cell cycle, and poly(A) RNA binding. Examples are MCM3, MCM4, MCM6, and MCM7 of the mini-chromosome maintenance protein complex, which regulates genomic DNA replication [26]. Another candidate found in

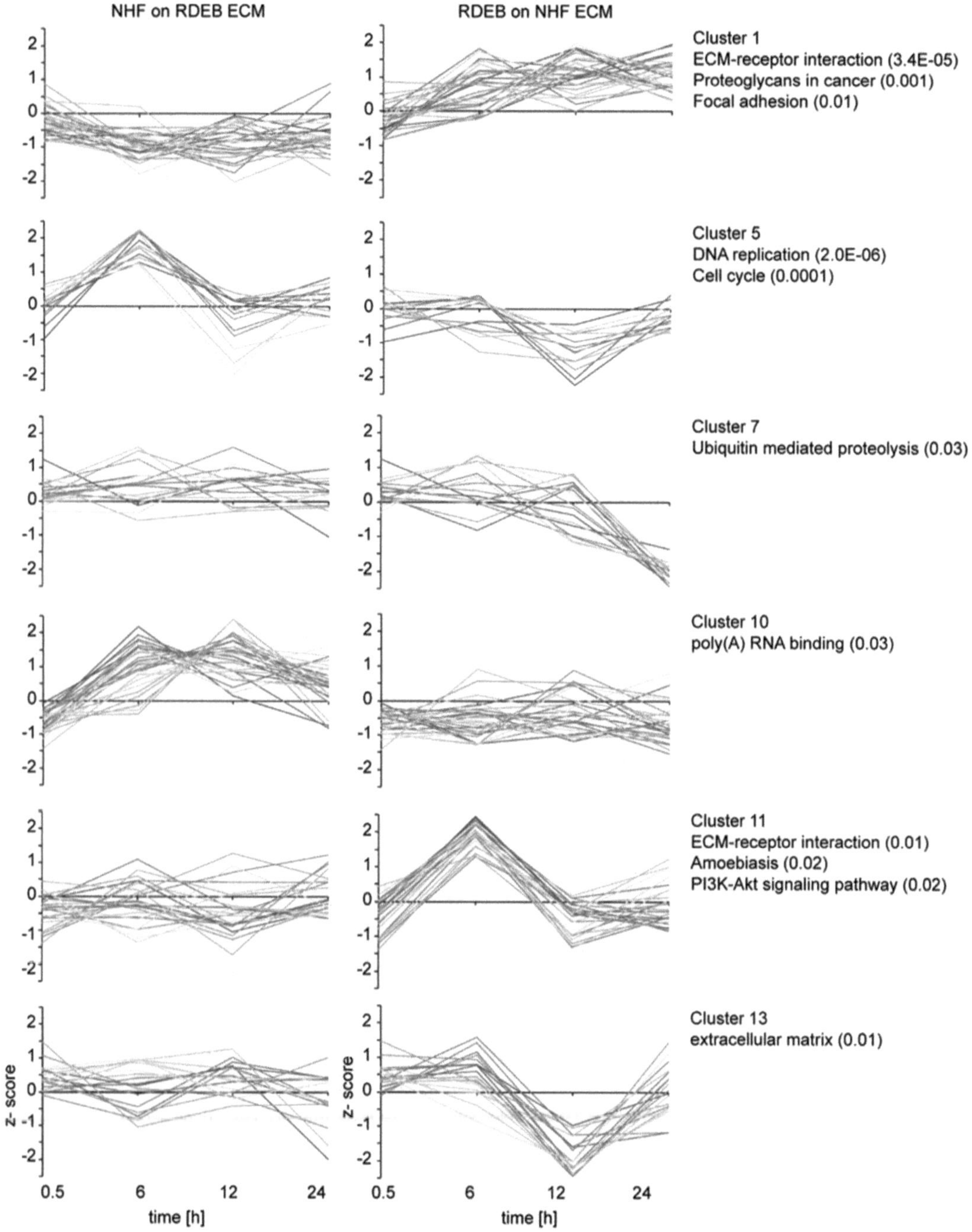

Fig. 3 K-means cluster analysis of significantly regulated proteins in NHF and RDEB fibroblasts (FDR: 0.05). Clusters were analyzed for enriched GO terms. Respective p-values are noted in brackets (Fisher's exact test, $p < 0.05$, BH corrected) [34]

this cluster is the dermal basement membrane protein extracellular matrix protein 1 (ECM1). ECM1 is part of suprastructures in the dermal-epidermal junction and thus regulates skin homeostasis [27, 28]. Its overexpression induces cell proliferation by activating EGFR and is linked to enhanced metastasis and poor prognosis in cancer [29–32]. Since RDEB patients develop aggressive squamous cell carcinomas [33], this might indicate a proliferation-supporting, pro-cancerogenic effect of collagen VII-negative RDEB ECM [10]. Thus, this experimental strategy reveals new insights into cell-ECM interactions and allows the discrimination between cell-intrinsic and ECM-specific influences on disease progression.

4 Notes

1. The third, light SILAC label is not used, as we use non-labeled ECM as a scaffold. With this approach, one can discriminate ECM proteins synthesized by reseeded, labeled cells from ECM proteins of the initially prepared matrix scaffold. Since the ECM is not used for data evaluation, it is generated in standard DMEM and not in SILAC DMEM.
2. Fibroblasts need proline for proper collagen fibril synthesis. The absence of proline will stimulate cells to convert heavy arginine into heavy proline, leading to protein quantification artifacts in MS data analysis. We add additional proline to reduce the arginine-to-proline conversion without interfering with the incorporation of heavy arginine. Also, an excess of arginine will enhance arginine-to-proline conversion. Thus, optimal arginine and proline concentrations should be titrated for each cell type.
3. dFBS is critical to ensure that only labeled variants of arginine and lysine are metabolized by cells.
4. Ascorbic acid is an essential cofactor of enzymes that catalyze proline and lysine hydroxylation [35], which in turn are critical for collagen stability. It can be prepared in a 5 mg/mL stock solution in dH_2O. The solution is sterile-filtered with a syringe-driven filter (0.2 μm, 30 mm diameter). The solution should be kept in the dark. Freeze stocks are stored at −20 °C.
5. Practicing the isolation of ECM in advance is advisable. Plates should be checked for efficacy under the microscope after each step. Backup plates for ECM isolation are desirable, in case the ECM is lost during the purification procedure. The suitable amount of produced ECM needs to be studied beforehand: thick ECM will easily detach from the culture dish.
6. The plug is shredded by centrifuging it through a nitrocellulose filter support grid. Dissolve the nitrocellulose filter membrane by acetone and wash the grid prior to shredding the gel plug. Place the grid in a new, clean tube and centrifuge the plug at maximum speed through the grid.

7. STAGE-tip syringes to stack C18 disks into a pipet tip can be built according to the online video instructions from the Max Planck Institute of Biochemistry (http://www.biochem.mpg.de/226863/Tutorials). The slurry volume of ReproSil-pure C18-basic depends on the amount of protein: samples with high protein concentrations require larger C18 volumes.
8. To identify false-positive hits, SILAC labels should be swapped between biological replicates.

Acknowledgments

We thank all members of the Protein Homeostasis laboratory of Prof. Jörn Dengjel for their help and discussions. We thank Prof. Leena Bruckner-Tuderman and the Molecular Dermatology laboratory of the University Clinic Freiburg i. Br. for providing the fibroblasts. This research was supported by the German Research Foundation (DFG) through grant DE 1757/3-2 and by the Swiss National Science Foundation (SNSF) through grant 31003A-166482/1.

References

1. Bonnans C, Chou J, Werb Z (2014) Remodelling the extracellular matrix in development and disease. Nat Rev Mol Cell Biol 15 (12):786–801
2. Figliuzzi M, Remuzzi G, Remuzzi A (2014) Renal bioengineering with scaffolds generated from rat and pig kidneys. Nephron Exp Nephrol 126(2):113
3. Shaikh FM et al (2008) Fibrin: a natural biodegradable scaffold in vascular tissue engineering. Cells Tissues Organs 188(4):333–346
4. Ross EA et al (2009) Embryonic stem cells proliferate and differentiate when seeded into kidney scaffolds. J Am Soc Nephrol 20(11):2338–2347
5. Has C et al (2018) Epidermolysis bullosa: molecular pathology of connective tissue components in the cutaneous basement membrane zone. Matrix Biol 71–72:313–329. https://doi.org/10.1016/j.matbio.2018.04.001
6. Burgeson RE (1993) Type VII collagen, anchoring fibrils, and epidermolysis bullosa. J Invest Dermatol 101(3):252–255
7. Becker AC et al (2012) Friend or food: different cues to the autophagosomal proteome. Autophagy 8(6):995–996
8. Fine J-D et al (2008) The classification of inherited epidermolysis bullosa (EB): Report of the Third International Consensus Meeting on Diagnosis and Classification of EB. J Am Acad Dermatol 58(6):931–950
9. Küttner V et al (2013) Global remodelling of cellular microenvironment due to loss of collagen VII. Mol Syst Biol 9:657
10. Mittapalli VR et al (2016) Injury-driven stiffening of the dermis expedites skin carcinoma progression. Cancer Res 76(4):940–951
11. Kuttner V et al (2014) Loss of collagen VII is associated with reduced transglutaminase 2 abundance and activity. J Invest Dermatol 134(9):2381–2389
12. Thriene K et al (2018) Combinatorial omics analysis reveals perturbed lysosomal homeostasis in collagen VII-deficient keratinocytes. Mol Cell Proteomics 17(4):565–579
13. Fischer R, Kessler BM (2015) Gel-aided sample preparation (GASP)--a simplified method for gel-assisted proteomic sample generation from protein extracts and intact cells. Proteomics 15(7):1224–1229
14. Gilar M et al (2005) Orthogonality of separation in two-dimensional liquid chromatography. Anal Chem 77(19):6426–6434
15. Batth TS, Francavilla C, Olsen JV (2014) Off-line high-pH reversed-phase fractionation for in-depth phosphoproteomics. J Proteome Res 13(12):6176–6186

16. Rappsilber J, Mann M, Ishihama Y (2007) Protocol for micro-purification, enrichment, pre-fractionation and storage of peptides for proteomics using StageTips. Nat Protoc 2 (8):1896–1906
17. Vlodavsky I (2001) Preparation of extracellular matrices produced by cultured corneal endothelial and PF-HR9 endodermal cells. Curr Protoc Cell Biol Chapter 10:Unit 10.4. https://doi.org/10.1002/0471143030.cb1004s01
18. Saveliev S et al (2013) Trypsin/Lys-C protease mix for enhanced protein mass spectrometry analysis. Nat Methods 10:1134
19. Rackiewicz M et al (2017) Hydrophobic interaction chromatography for bottom-up proteomics analysis of single proteins and protein complexes. J Proteome Res 16(6):2318–2323
20. Tyanova S et al (2016) The Perseus computational platform for comprehensive analysis of (prote)omics data. Nat Methods 13 (9):731–740
21. Naba A et al (2016) The extracellular matrix: tools and insights for the "omics" era. Matrix Biol 49:10–24
22. Saeed AI et al (2003) TM4: a free, open-source system for microarray data management and analysis. BioTechniques 34(2):374–378
23. Szklarczyk D et al (2015) STRING v10: protein-protein interaction networks, integrated over the tree of life. Nucleic Acids Res 43(Database issue):D447–D452
24. Nystrom A et al (2015) Losartan ameliorates dystrophic epidermolysis bullosa and uncovers new disease mechanisms. EMBO Mol Med 7 (9):1211–1228
25. Nystrom A et al (2013) Collagen VII plays a dual role in wound healing. J Clin Invest 123 (8):3498–3509
26. Neves H, Kwok HF (2017) In sickness and in health: the many roles of the minichromosome maintenance proteins. Biochim Biophys Acta 1868(1):295–308
27. Chan I (2004) The role of extracellular matrix protein 1 in human skin. Clin Exp Dermatol 29 (1):52–56
28. Sercu S et al (2008) Interaction of extracellular matrix protein 1 with extracellular matrix components: ECM1 is a basement membrane protein of the skin. J Invest Dermatol 128(6):1397–1408
29. Lee KM et al (2014) Extracellular matrix protein 1 regulates cell proliferation and trastuzumab resistance through activation of epidermal growth factor signaling. Breast Cancer Res 16 (6):479
30. Lee KM et al (2015) ECM1 regulates tumor metastasis and CSC-like property through stabilization of beta-catenin. Oncogene 34 (50):6055–6065
31. Han Z et al (2001) Extracellular matrix protein 1 (ECM1) has angiogenic properties and is expressed by breast tumor cells. FASEB J 15 (6):988–994
32. Wang L et al (2003) Extracellular matrix protein 1 (ECM1) is over-expressed in malignant epithelial tumors. Cancer Lett 200(1):57–67
33. Fine JD et al (2009) Epidermolysis bullosa and the risk of life-threatening cancers: the National EB Registry experience, 1986–2006. J Am Acad Dermatol 60(2):203–211
34. Szklarczyk D et al (2017) The STRING database in 2017: quality-controlled protein-protein association networks, made broadly accessible. Nucleic Acids Res 45(D1): D362–D368
35. Murad S et al (1981) Regulation of collagen synthesis by ascorbic acid. Proc Natl Acad Sci U S A 78(5):2879–2882

Chapter 16

Standard Preparation Protocol of Human Skin Samples for Transmission Electron Microscopy

Gery Barmettler and Urs Ziegler

Abstract

This chapter is intended to be a guide to skin studies dealing with the analysis of subcellular structures of cells and tissues by providing an exhaustive sample preparation protocol for transmission electron microscopy (TEM). Critical steps to obtain good ultrastructure are mentioned as well as notes to improve the sample collection. New ways to automatize the imaging acquisition are discussed.

Key words TEM tissue preparation, Skin tissues, Skin substitutes, Bioengineered skin analogs, Tissue-engineered dermo-epidermal skin, Human-pigmented skin analog, Mesenchymal cells, Epidermal stratification and cornification tissue engineering

1 Introduction

The classical TEM analyses are still the standard methods for the representation of the ultrastructure of medical or biological preparations per se. In contrast to light microscopes, electron microscopes have a much higher resolution. This allows to resolve structural details and the morphology of subcellular components of cells, such as membranes, ribosomes, Golgi apparatus, cytoskeleton, mitochondria, vesicles, cell nuclei, and cell-cell contacts. Such structural detail can only be observed by electron microscopy in optimally prepared samples. This wealth of information in structural details is also the reason why TEM techniques remain essential today as well as in the future.

This method is routinely used in many areas of medicine and life sciences, so it is essential that the preservation of the ultrastructure is a prerequisite for the morphological evaluation of the data. One of the most crucial steps for analysis by TEM is the fixation of the native sample. Fixation must aim to preserve the ultrastructure as closely as possible to the native state. Except for cryo-electron microscopy, fixation is done using chemical reactions to stabilize

Sophie Böttcher-Haberzeth and Thomas Biedermann (eds.), *Skin Tissue Engineering: Methods and Protocols*, Methods in Molecular Biology, vol. 1993, https://doi.org/10.1007/978-1-4939-9473-1_16,

the structure. To obtain reproducible results the concentrations of buffers and fixatives, such as glutaraldehyde (GA) and osmium tetroxide (Os), the quality of reagents, as well as the reaction time must be precisely controlled [1, 2].

An alternative to classical chemical fixation is fixation by high-pressure freezing (HPF). HPF results in a better ultrastructure preservation than the chemical fixation. However, one of the limitations of this method is the size of the specimen which should not be thicker than around 0.2 mm [3, 4].

The following protocol is based on standard preparations performed at the Center for Microscopy and Image Analysis of the University of Zurich [5–9]. Notes at the end give precise and further details on the preparation steps below.

2 Materials

2.1 Chemicals

1. Sodium cacodylate powder.
2. Uranyl acetate (UAc) powder.
3. Lead(II) nitrate.
4. Lead(II) citrate.
5. Sodium hydroxide pellets.
6. Epon 812.
7. Durcupan ACM.
8. Dibutylphthalate.
9. Dodecenylsuccinic anhydride (DDSA).
10. Accelerator DMP-30.
11. Molecular sieve dehydrate.
12. Ethanol absolute.
13. Propylene oxide.
14. 1 g Osmium tetroxide in ampoule.

2.2 Stock Solutions

1. 25% Glutaraldehyde (GA) solution, EM grade.
2. 2% Osmium tetroxide (Os) aqueous solution.
3. 0.2 M Sodium cacodylate buffer, pH 7.35, in water.
4. 3% Uranyl acetate in water.
5. Epon-Araldite stock solution (without accelerator DMP-30, for approximately 150 mL): 70.89 g Epon 812, 92.35 g Durcupan ACM, 8.68 g Dibutylphthalate.

2.3 Working Solutions

1. 1% Formvar solution in ethylene dichloride.
2. GA fixation buffer: 2.5% GA in 0.1 M sodium cacodylate buffer.

3. Rinsing buffer: 0.1 M Sodium cacodylate buffer.
4. 1% Uranyl acetate in water.
5. Os fixation buffer: 1% (w/v) Osmium tetroxide in 0.1 M sodium cacodylate buffer.
6. UAc staining solution: 1% UAc in water.
7. Epon-Araldite working solution (approx. 20 mL): 11.7 g Epon-Araldite stock solution, 10.0 g DDSA, 620 mg accelerator.
8. 1% (w/v) Toluidine blue in water.
9. Reynolds' lead solution [10]: (A) 1.33 g Lead(II) nitrate in 15 mL water; (B) 1.76 lead(II) citrate in 15 mL water; (C) 8 mL 1 N NaOH. Mix A and B into a 50 mL volumetric flask, and shake well until solution becomes homogeneously milky. Add 8 mL of solution C. Solution becomes clear. Fill up to 50 mL with water. The finished solution can be portioned and stored at 4 °C for several months.
10. 3% UAc in water.

2.4 Materials

1. Silicon nitride membrane, membrane 50 nm, 1 mm × 1 mm.
2. Copper grid 2 mm × 1 mm.
3. Tabbed copper grid.
4. Single-edge blades.
5. Double-edge stainless steel razor blade.
6. Filter paper, hardened ashless.
7. Parafilm®.

2.5 Instruments

1. Water system for 18.2 MΩ cm ultrapure water.
2. Diamond knives for ultrathin sectioning.
3. Diamond trimming knives.
4. Ultramicrotome.

3 Methods

1. Extract the sample and immerse it immediately in the GA fixation buffer (*see* **Notes 1–8**).
2. Subsequently, cut samples with razor blades to a size of 1 mm in one dimension. Keep the sample immersed always in GA fixation buffer (Fig. 1).
3. Incubate 1 mm^3 sample in 1 mL GA fixation buffer for 2–4 h or overnight at 4 °C.
4. Wash 2× with 1 mL of rinsing buffer for 5 min each.
5. Postfix with 1 mL Os fixation buffer on ice (*see* **Note 9**) for 1 h.

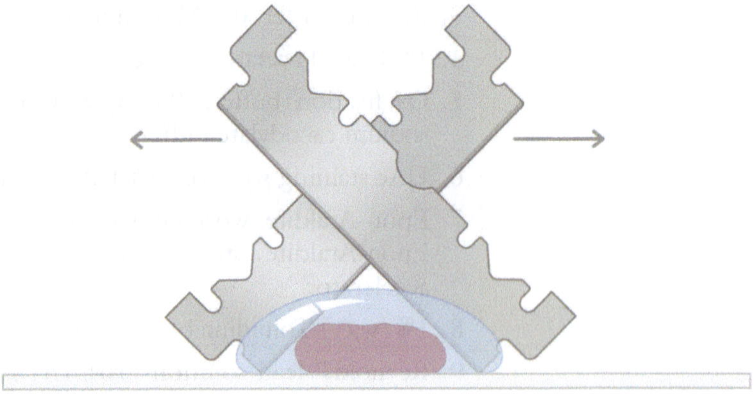

Fig. 1 Cutting tissue with two halves of a razor blade immersed in GA fixation buffer to reduce compression artifacts

6. Wash 3× with 1 mL water at room temperature (RT) for 5 min each.
7. Stain with UAc staining solution (*en bloc* staining) for 1 h at RT or overnight at 4 °C (*see* **Note 10**).
8. Wash 2× with water for 5 min each at RT.
9. Make a dehydration by an ascending ethanol series: 20 min 70% (v/v) ethanol, 20 min 80% (v/v) ethanol, and 100% dried ethanol 2× for 20 min each (*see* **Notes 11** and **12**).
10. 2× Propylene oxide for 20 min each (*see* **Note 13**).
11. 2× 1:1 Propylene oxide/Epon-Araldite working solution for 1 h each, or 1× overnight.
12. 1× Epon-Araldite working solution for 1 h.
13. Embed in fresh Epon-Araldite working solution in a silicon mold (*see* **Note 14**).
14. Polymerization for about 24 h at 60 °C.
15. Trimming, semi-thin and ultrathin sectioning (*see* **Notes 15–19**).
16. Collection of the ultrathin sections (*see* Fig. 2 and **Note 20**).
17. Post-staining (*see* **Notes 21** and **22**).
18. Imaging at transmission electron microscope (*see* **Note 23**).

4 Notes

1. The quality of the water for all the buffers and the solutions in electron microscopy technique is an essential factor. New water systems produce 18.2 MΩ·cm ultrapure water which is recommended to use.
2. The optimal concentration of the fixative and the best buffer must be determined if other buffers are employed or other

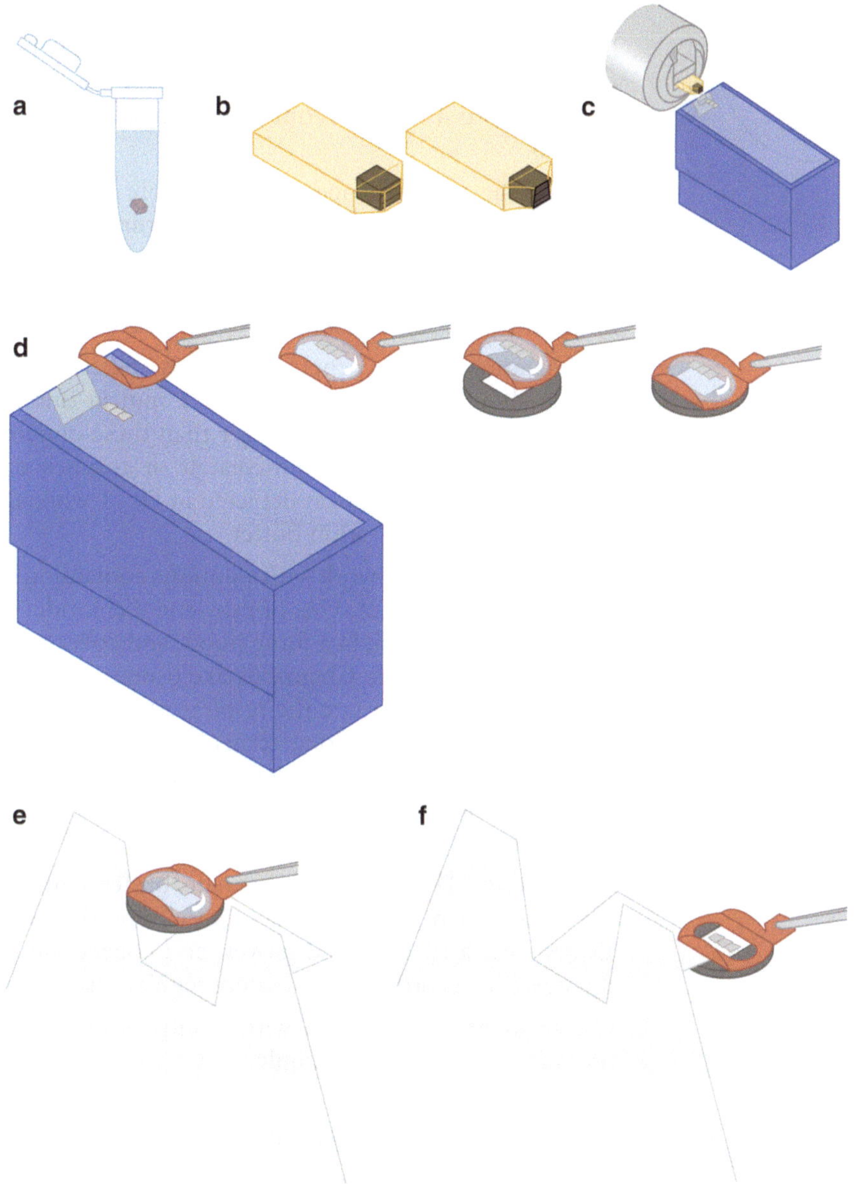

Fig. 2 Illustration showing the TEM sample preparation steps. (**a**) Process the tissue in an Eppendorf tube: first fixation, washing; second fixation, washing; *en bloc* staining, washing, dehydration, Epon-Araldite. In case of a high number of samples, a processing device (Leica EM TP) is recommended. (**b**) Embed the tissue in Epon-Araldite in a silicon mold and bring the tissue in the position with the help of a toothpick under a binocular as long as the Epon-Araldite is liquid. After hardening in an oven at 60 °C for 24 h, first trim the block by hand with a razor blade, followed by a fine trimming using a glass or a diamond trimming knife (Diatome) in an ultra-sectioning device (Reichert or Leica). (**c**) Ultra-sectioning with a diamond knife (Diatome), 60–80 nm sections. (**d**) Pick up ultra-sections with a self-made modified slot grid (*see* Fig. 3) and mount them on a silicon nitride support film or a formvar-coated slot grid. (**e**) Under a binocular, suck away the water very slowly, selectively, and alternately with the filter paper. (**f**) Separate the two with the filter paper

samples than skin tissue are processed. We recommend for human skin tissue 2.5% glutaraldehyde (GA) in 0.1 M sodium cacodylate buffer (pH 7.35). PBS has a very low buffer capacity and therefore it is not recommended as a buffer solution for fixation.

3. It is of the utmost importance to apply a freshly prepared fixative solution and at the same temperature as the tissue itself. Additionally, it must be applied immediately and without delay. Best results are achieved if the fixatives react with the biological material within seconds after a biopsy is taken [2].
4. To guarantee proper penetration of the fixation solution the sample should not be thicker than 1 mm in one dimension [2]. If extracted specimen is larger than these dimensions, cut into smaller pieces either immediately or as quick as possible, ideally using razor blades as depicted in Fig. 1 while the sample is immersed in GA fixation buffer.
5. CAUTION: For the work with solutions containing hazardous chemicals like osmium, GA, or ethylene dichloride, wear a lab coat and gloves, work in a fume hood, and collect the waste in labeled containments. Chemicals like osmium, GA, and ethylene dichloride can irritate the respiratory tract, eyes, and skin, and can cause chronic (long-term) health effects. Waste of all these substances must be collected and disposed according to the guidelines of Hazardous Waste Management of the University.
6. CAUTION: The sodium cacodylate buffer contains arsenic which is carcinogen and may cause reproductive damage; always wear a lab coat and gloves, and collect and dispose of the waste in accordance with safety regulations.
7. Use fresh fixative solution with a surplus of approx. 10–20 times the volume of the sample.
8. Prevent any drying/pressing/squeezing or other damage of the sample as it may lead to degradation of the ultrastructure.
9. After washing of the sample with the buffer twice, it is optimal to make a second fixation with Os fixation solution in water. Osmium is a heavy metal and binds to the lipid membranes; this later provides a stronger contrast of these structures.
10. The *en bloc* staining technique with 1% UAc enhances the contrast of the tissue [11]. It is possible to omit this step and stain with a 2% uranyl acetate solution in water at the end of the procedure, onto the ultrathin sections for 5–20 min. But, the *en bloc* staining with 1% UAc generates a stronger contrast due to the diffusion of UAc through the tissue and the binding in the tissue is more powerful than the binding just on the surface of the ultrathin sections.

11. Do not use denatured ethanol. Effects of denaturants on tissue are not known.
12. Use molecular sieve with a size of 3 Å for the 100% ethanol to produce water-free ethanol. A high water content in the sample can result in soft resin and thus difficulty in sectioning.
13. Propylene oxide has a high gas pressure. Using safe-lock Eppendorf tubes or sealing the tube with a strip of Parafilm is recommended to prevent evaporation.
14. Once the samples are immersed in Epon-Araldite working solution use a toothpick under a binocular to align the samples that they will later be correctly positioned for ultrathin cutting (Fig. 2c).
15. CAUTION: Epon-Araldite contains hazardous substances. Embedding must be carried out in a fume hood and wear gloves dense and resistant against solvents (Sempercare® nitrile skin[2] are recommended). The waste must be collected and disposed strictly in accordance with safety regulations.
16. Trimming to trapezoidal or rectangular pyramids is preferred. Sharp edges ensure better pressure distribution during cutting, which allows qualitatively better sections.
17. Semi-thin sections are optionally used to facilitate the correct localization of the target area in the tissue. For this purpose, 100–200 nm sections are made with the diamond knife (Diatome 35° or 45°). These sections are applied to a glass slide by a large, stainless steel loop and dried on a heating plate at 50–60 °C. Drops of filtered toluidine blue 1% (w/v) in water are applied for 1–5 min, depending on tissue. Afterwards, the microscope slide is rinsed with water and air-dried. Locate the region of interest using phase-contrast light microscopy. Subsequently trim block around identified target area.
18. A critical step to perform in automatic imaging is to mount the sections as flat as possible on support materials. Silicon nitride membranes are atomically flat, are chemically inert, do not deform in the TEM beam, have no type of impurities, and are available in ready-to-use units. For many samples, the silicon nitride membrane (Ted Pella, Plano) is recommended. For good section pickup, a self-made shaped unit is first produced in a very simple way with a tabbed copper slot grid. *See* Fig. 3.
19. The ultrathin cut can now be carried out. The diamond knife is used to make 60–80 nm cuts. The thickness of the sections can be estimated according to the optical interference color [12]: gray 50–60 nm, silver 60–90 nm, gold 90–150 nm, and purple 150–190 nm. According to this color scale, the ultramicrotome is adjusted. Once the sections are produced they float on the water surface and are manipulated with a hair on a stick (wooden or plastic) in the direction that the sections can be captured with an empty slot grid.

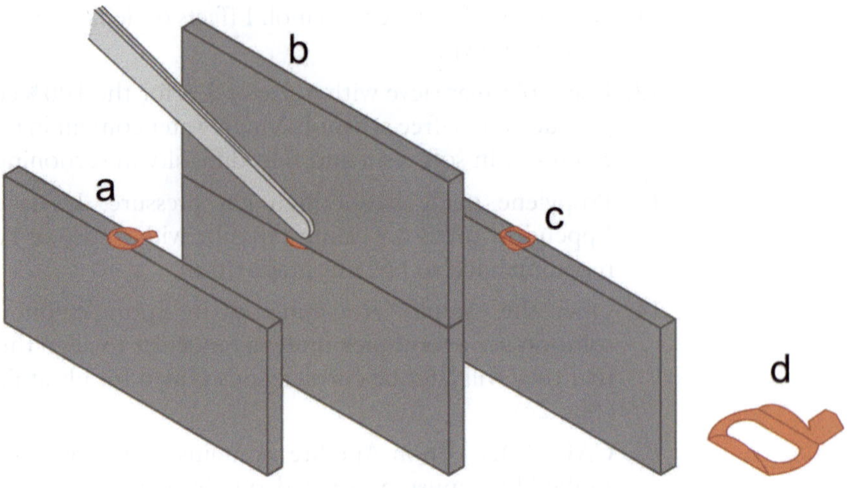

Fig. 3 Produce a perfect eyelet for section collection: a brass or aluminum plate with a thickness of 2 mm is used. (**a**) Place the tabbed copper slot grid on the edge. (**b**) Take a second, same plate and place it on top of it. Fasten the two plates with fingers or clamps. Form with the backside of a tweezer the protruding parts of the tabbed copper slot grid over the edges. (**c**) Release the plates. (**d**) Remove the eyelet from the plate; it is now ready to use

20. Wrinkle-free cutting on the grid is a goal that can be mastered with the following method: in addition to the usual cutting tools, a binocular, a Petri dish with filter paper, and the self-formed slotted grid (Fig. 3d) are required. Cut the filter paper into a form (Fig. 2e) and glue it into the Petri dish with double-sided adhesive tape. The sections float on the water surface of the diamond knife and are picked up with a tweezer using the self-formed slot grid. By pressing lightly on the water surface, the sections get stuck in the small drop. Then pull very close to the silicon nitride membrane and let only the small water droplet on the underside touch. Both grids now stick together. Under the binoculars, the water between them is sucked off very slowly, selectively, and alternately at the prepared filter paper (Fig. 2e) until only a little water remains visible. The small water cushion reduces mechanical effects on the surface. Then pull the two over the edge of the filter paper and pull the filter paper through between the two (Fig. 2f). When properly moved, the silicone membrane falls onto the filter paper in the Petri dish. The grid can now be placed in the grid box with tweezers until further processing.
21. First post-staining: Put grid on one drop of UAc 1% in water for 5–20 min, short wash on three drops of water, and suck water away with a filter paper. Remarks: Post-staining is recommended to enhance contrast.
22. Second post-staining: Put gird on one drop of Reynolds' lead solution [10] for 5 min, short wash on three drops of water,

and suck water away with a filter paper. Store grid in a grid box. Remarks: A few sodium hydroxide pellets are put around this drop. These prevent precipitates, by lead solution reacting with CO_2 from the air. Remarks: Uranyl acetate and lead solution should be centrifuged before use in an Eppendorf tube at $14,000 \times g$ for 3 min. Staining should be performed in a closed Petri dish on a Parafilm, which is covered for incubation. UAc and lead solutions must be carefully collected and disposed according to safety regulations.

The sample is now ready for microscopy.

23. Once samples are processed according to the protocol described above ultrathin sections are examined with a transmission electron microscope (optionally, a short post-staining step can be performed). Figure 4 shows representative TEM images of different skin areas acquired by an investigator

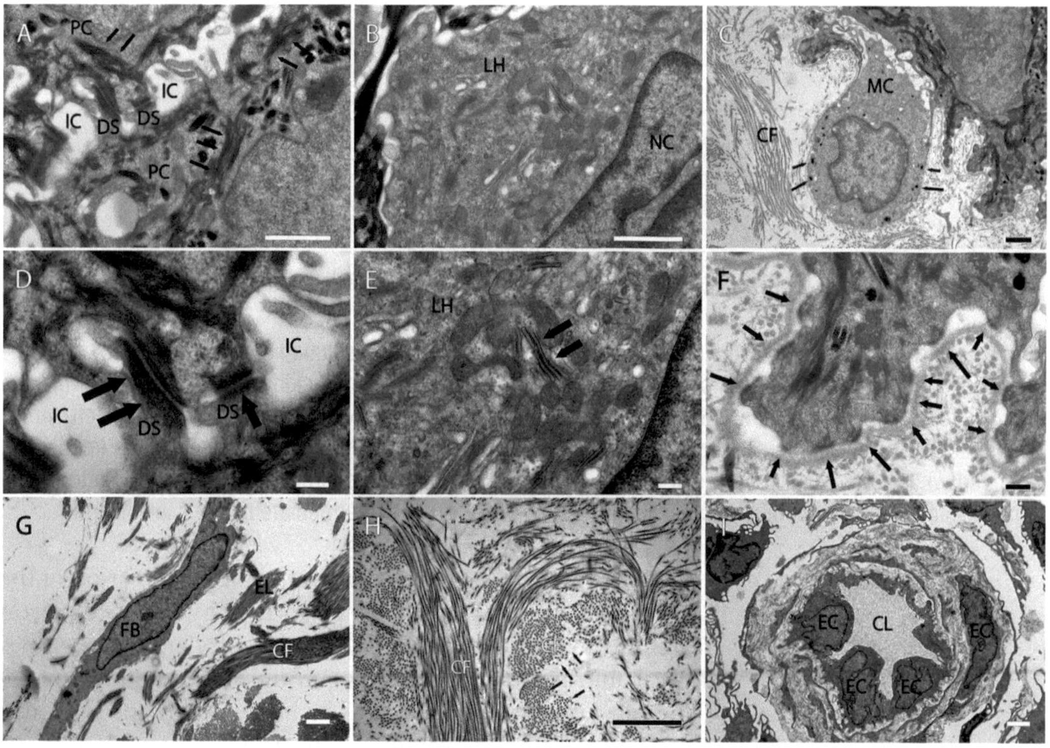

Fig. 4 Transmission electron microscopy images of human skin sample: (**a**) In the layer of the stratum spinosum: prickle cells (PC) with tonofilaments (arrows) and desmosomes (DS), in between of the two prickle cells (PC) intercellular space (IC). (**b**) Dendritic cell (Langerhans cell cytoplasm (LH), nucleus (NC). (**c**) In the layer of the stratum basale: melanocyte (MC) with melanin granules (arrows), collagen fibers (CF). (**d**) Same region as A at higher magnification desmosomes is well observed (arrows). (**e**) On higher magnification, same region as B, Langerhans cell (LH) with Birbeck bodies (arrows). (**f**) Same region as C, hemidesmosomes (arrows) anchored at the basal lamina. (**g–i**) Dermis **G** Fibroblast (FB), collagen fibers (CG), elastin (EL). (**h**) Collagen fibers longitudinally sectioned (CG), collagen fibers cross-sectioned (arrows). (**i**) Capillary with endothelial cells (EC), capillary lumen (CL). Scales: **a–c** 1 μm, **d–f** 200 nm, **g–i** 2 μm

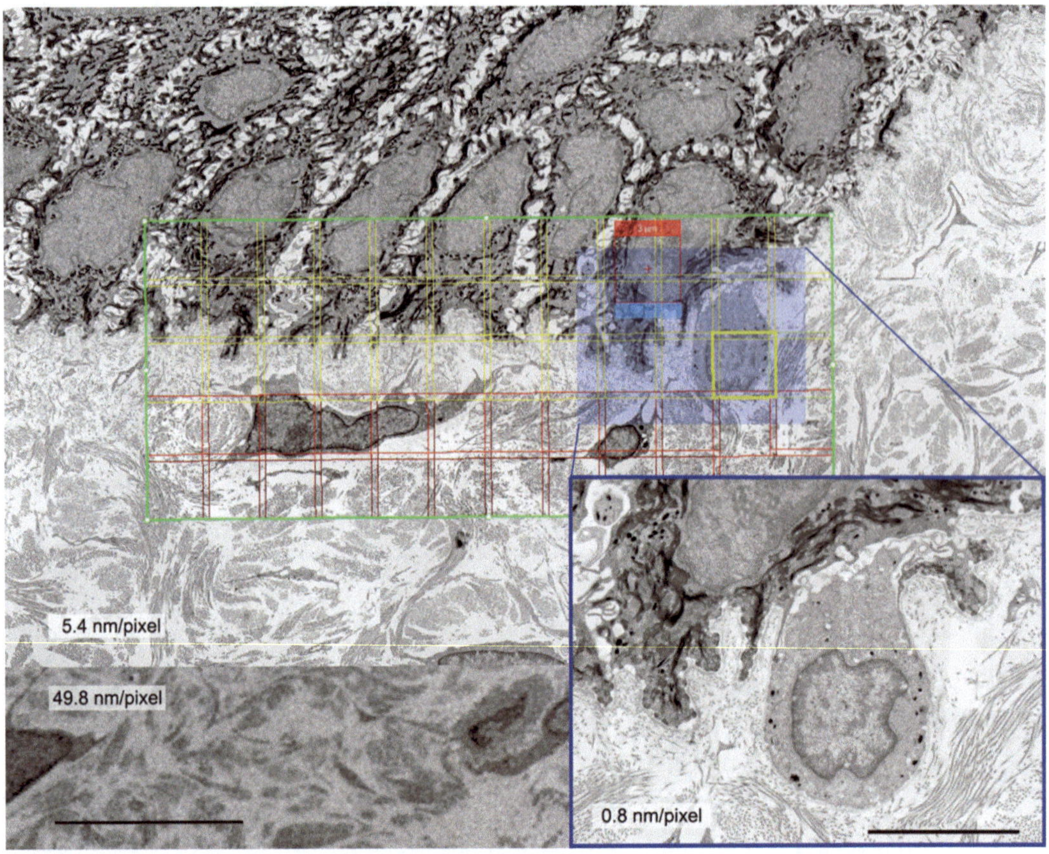

Fig. 5 TEM imaging using an overview approach with selected higher resolution areas of interest of the same human skin tissue: an ultrathin cross section, 70 nm, of a normal human skin is seen. The epidermis and dermis are shown. The overview of the complete section is created first with 49.8 nm/pixel (bottom left). Areas of interest are selected and acquired at intermediate resolution of 5.4 nm/pixel. High-resolution images are acquired with 0.8 nm/pixel; see green rectangular area with smaller squares, red and yellow. Bottom right image: stitched images of blue area, scale bar 4 μm. Overview and intermediate image scale 10 μm

selecting manually the areas of interest. The examination of the samples at the TEM is normally performed by a trained person searching for the structures of interest. Modern microscopes are capable of automatically acquiring and stitching images (Thermo Fisher Talos 120 with the integrated Maps software) allowing to obtain an overview of the grid. The investigator can draw areas of interest in overview images and acquire the selected area of interests at higher resolution subsequently (*see* Fig. 5). This multiscale and panoramic approach will help to better understand the high-resolution images, to locate them in the context of the tissue, and will reduce artifacts by otherwise looking at very small parts of the sample. The interpretations of the results are easier, quicker, and less biased because of the larger context information contained in the overview.

Acknowledgments

Many thanks to Drs. José María Mateos and Andres Kaech for critically reading of the manuscript, thoughtful comments, and many advices. Special thanks to my former colleague Dr. Bruno Guhl, who trained me with his outstanding craftsmanship about the techniques of electron microscopy. I thank the whole team of the Center for Microscopy and Image Analysis for the excellent collaboration over the years. Illustrations: University of Zurich, Information Technology, MELS/SIVIC, Nadja Baltensweiler.

References

1. Hayat MA (1970) Principles and techniques of electron microscopy. Van Nostrand Reinhold Company, New York
2. Griffiths G (1993) Fixation for fine structure preservation and immunocytochemistry. In: Griffiths G (ed) Fine structure immunocytochemistry. Springer, Berlin Heidelberg. https://doi.org/10.1007/978-3-642-77095-1_3
3. Kaech A et al (2008) Bal-Tec Hpm 010 high-pressure freezing machine. In: Cavalier A, Spehner DS, Humbel BM (eds) Handbook of cryo-preparation methods for electron microscopy. CRC Press, Boca Raton
4. Kaech A, Ziegler U (2014) High-pressure freezing: current state and future prospects. Methods Mol Biol 1117:151–171
5. Pontiggia L et al (2014) De novo epidermal regeneration using human eccrine sweat gland cells: higher competence of secretory over absorptive cells. J Invest Dermatol 134(6):1735–1742
6. Biedermann T et al (2014) Tissue-engineered dermo-epidermal skin analogs exhibit de novo formation of a near natural neurovascular link 10 weeks after transplantation. Pediatr Surg Int 30(2):165–172
7. Biedermann T et al (2015) The influence of stromal cells on the pigmentation of tissue-engineered dermo-epidermal skin grafts. Tissue Eng Part A 21(5–6):960–969
8. Böttcher-Haberzeth S et al (2013) 'Trooping the color': restoring the original donor skin color by addition of melanocytes to bioengineered skin analogs. Pediatr Surg Int 29 (3):239–247
9. Böttcher-Haberzeth S et al (2013) Human eccrine sweat gland cells turn into melanin-uptaking keratinocytes in dermo-epidermal skin substitutes. J Invest Dermatol 133 (2):316–324
10. Reynolds ES (1963) The use of lead citrate at high pH as an electron-opaque stain in electron microscopy. J Cell Biol 17:208–212
11. Maunsbach AB, Afzelius BA (1999) Biomedical electron microscopy. Academic Press, Cambridge
12. Peachey LD (1960) Section thickness and compression. In: Bargmann W, Möllenstedt G, Niehrs H, Peters D, Ruska E, Wolpers C (eds) Vierter Internationaler Kongress für Elektronenmikroskopie/Fourth International Conference on Electron Microscopy/Quatrième Congrès International de Microscopie Électronique. Springer, Berlin, Heidelberg

Chapter 17

Methods for Assessing Scaffold Vascularization In Vivo

Jiang-Hui Wang, Jinying Chen, Shyh-Ming Kuo, Geraldine M. Mitchell, Shiang Y. Lim, and Guei-Sheung Liu

Abstract

The success of tissue engineering hinges on the rapid and sufficient vascularization of the neotissue. For efficient vascular network formation within three-dimensional (3D) constructs, biomaterial scaffolds that can support survival of endothelial cells as well as formation and maturation of a capillary network in vivo are highly sought after. Here, we outline a method to biofabricate 3D porous collagen scaffolds that can support extrinsic and intrinsic vascularization using two different in vivo animal models—the mouse subcutaneous implant model (extrinsic vascularization, capillary growth within the scaffold originating from host tissues outside the scaffold) and the rat tissue engineering chamber model (intrinsic vascularization, capillary growth within the scaffold derived from a centrally positioned vascular pedicle). These in vivo vascular tissue engineering approaches hold a great promise for the generation of clinically viable vascularized constructs. Moreover, the 3D collagen scaffolds can also be employed for 3D cell culture and for in vivo delivery of growth factors and cells.

Key words Porous collagen scaffolds, Extrinsic vascularization, Intrinsic vascularization, In vivo model

1 Introduction

Porous scaffolds have been widely used in the application of tissue-regenerative medicine to provide a physical microenvironment to guide the formation of new tissues and organs. In particular, collagen-based porous scaffolds have been successfully used as cell-carrying vehicles in many clinical applications for tissue repair and regeneration due to the biological inducibility and desirable mechanical and degradable properties of collagen [1, 2]. Collagen is one of the most abundant structural proteins that maintains the biological and structural integrity of extracellular matrix architecture [3]. Scaffolds made of collagen can provide a temporary

Jiang-Hui Wang, Jinying Chen, Shiang Y. Lim, and Guei-Sheung Liu have contributed equally to this work. Shiang Y. Lim and Guei-Sheung Liu should be regarded as equal senior authors.

Sophie Böttcher-Haberzeth and Thomas Biedermann (eds.), *Skin Tissue Engineering: Methods and Protocols*, Methods in Molecular Biology, vol. 1993, https://doi.org/10.1007/978-1-4939-9473-1_17,

mechanical support for tissue growth and provide a biomimetic microenvironment for vascularization, a fundamental biological process which is indispensable for survival of scaffold-implanted cells and development of functionally viable tissue constructs. We have recently demonstrated a type I bovine collagen-derived three-dimensional (3D) porous collagen scaffold with bioactivity that mimics the functions of natural extracellular matrix to support vascularization in vivo in different experimental models [4].

In this chapter, we outline the steps required for preparation of collagen porous scaffolds. We also describe two experimental models, the mouse subcutaneous implant model [4] and rat tissue engineering chamber model [4, 5], to assess the potential of the collagen scaffolds in promoting vascularization in vivo. This methodology can be readily adapted as an assay model for studies on angiogenesis, 3D cell culture, and delivery of growth factors and cells in vivo.

2 Materials

2.1 Porous Collagen Scaffold Preparation

1. Type I collagen (isolated from cow skin).
2. Pepsin (Sigma-Aldrich, MO, USA).
3. Acetic acid (Merck Millipore, Darmstadt, Germany): 0.5 N in dH_2O.
4. Sodium chloride (NaCl; Sigma-Aldrich): 2.5 M in dH_2O.
5. Tris base: 50 mM Tris(hydroxymethyl)aminomethane (Sigma-Aldrich) in dH_2O.
6. The NuPAGE® Precast Gel System (Thermo Fisher Scientific, VIC, Australia).
7. COL1A1 (bovine collagen, type I) ELISA kit (MyBioSource, CA, USA).
8. N-ethyl-N′-(3-dimethylaminopropyl) carbodiimide (Sigma-Aldrich): 25 mM in 80/20 acetone/PBS.
9. N-hydroxysuccinimide (Sigma-Aldrich): 12.5 mM in 80/20 acetone/PBS.
10. Phosphate-buffered saline (PBS; pH 7.4): 137 mM NaCl, 2.7 mM KCl, 10 mM Na_2HPO_4, and 1.8 mM KH_2PO_4 in dH_2O.
11. Material-testing system (MTS Systems Corporation; MN, USA).
12. Vernier caliper.
13. Ethanol (Sigma-Aldrich): $\geq$99.8%.

2.2 Mouse Subcutaneous Scaffold Implantation

1. Male C57B/L6 mice (10–12 weeks).
2. Chlorhexidine (Sigma-Aldrich): 0.5% in 70% ethanol.
3. Stainless steel wound clip.

4. Carprofen (Rimadyl).
5. Pentobarbital.

2.3 Rat Tissue Engineering Chamber

1. Male Sprague-Dawley rat (300–400 g).
2. Chlorhexidine (Sigma-Aldrich): 0.5% in 70% ethanol.
3. 6-0 Polypropylene suture (Ethicon, NJ, USA).
4. 4-0 silk suture (Ethicon).
5. Polyacrylic chamber (internal dimensions of $10 \times 8 \times 4$ mm; Department of Chemical and Biomolecular Engineering, University of Melbourne, VIC, Australia).
6. Human fibrinogen (Sigma-Aldrich): 15 mg/mL in dH_2O.
7. Human thrombin (Sigma-Aldrich): 25 U/mL in dH_2O.
8. Carprofen.
9. Lethobarb.

2.4 Histological and and Immunohistochemical Assessment of Tissue Vascularization

1. Paraformaldehyde (PFA; Sigma-Aldrich): 4% in dH_2O.
2. Ethanol (Sigma-Aldrich): ≥99.8%.
3. Histolene (Thermo Fisher Scientific).
4. Paraffin wax.
5. Hematoxylin and eosin staining solution (Sigma-Aldrich).
6. Hydrogen peroxide solution (H_2O_2; Sigma-Aldrich): 30% in dH_2O.
7. Proteinase K (pH 7.8; Dako, Hamburg, Germany).
8. Protein block solution (Dako).
9. Rat anti-mouse CD31 (BD Biosciences, North Ryde, Australia).
10. Biotinylated rabbit anti-rat IgG (Dako).
11. Biotinylated Griffonia Simplicifolia lectin (B-1105; Vector Laboratories, Peterborough, UK).
12. Vectastain Elite ABC kit (Vector Laboratories).
13. Horseradish peroxidase-streptavidin (Dako).
14. Diaminobenzidine chromogen (Dako).
15. DPX mounting medium (VWR International, Poole,UK).
16. Video microscope with a computer-assisted stereo investigator system (MBF Bioscience, VT, USA).

3 Methods

3.1 Biofabrication of the Porous Collagen Scaffold

1. Prepare the freeze-dried collagen reagent for the generation of the porous collagen scaffolds with an average pore diameter of 80 μm.

2. Cut cow skin into small pieces of about 5 mm^3 and thereafter rinse with distilled water.
3. Soak the cut cow skin in 0.5 N acetic acid solution for 48 h.
4. Add homogenized cow skin into 0.5 N acetic acid solution containing 0.5 mg/mL pepsin and incubate at 4 °C for 24 h.
5. Centrifuge at (14,822 × *g*) for 1 h and collect the supernatant.
6. Add 2.5 M NaCl to precipitate type I collagen and centrifuge at 14,822 × *g* for 1 h.
7. Remove the supernatant and resuspend the type I collagen in 0.05 N acetic acid solution.
8. Dialyze type I collagen solution twice with 50 mM Tris base/1 M NaCl (pH 7.2) and then PBS.
9. Characterize type I collagen by SDS-polyacrylamide gel electrophoresis and determine the concentration by using ELISA.
10. Adjust the final concentration of the collagen solution to 7 mg/mL with PBS solution.
11. Freeze the collagen solution at −20 °C for 8 h, and then subject to a lyophilization procedure for 24 h to produce an interconnected porous structure.
12. Soak the semi-manufactured scaffolds in a cross-linking solution of 25 mM 1-ethyl-3-(3-dimethylaminopropyl) carbodiimide and 12.5 mM N-hydroxysuccinimide in an 80/20 acetone/PBS solution for 4 h and then wash twice with PBS.
13. Lyophilize the collagen scaffolds to enhance the mechanical intensity.
14. Image the collagen scaffolds by a scanning electron microscope for their 3D architecture (Fig. 1).

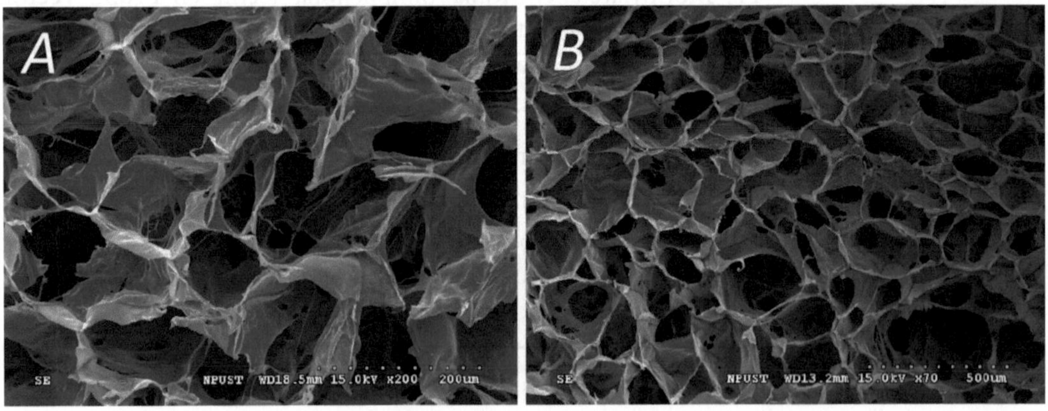

Fig. 1 The microstructures of the porous collagen scaffold. Representative images captured by scanning electron microscopy on the surface (**A**) and through a cross section (**B**) of a collagen scaffold

3.2 Characterization of the Mechanical Properties (Mechanical Strength, Porosity, and Water Content) of the Collagen Scaffold

1. Soak the collagen scaffold in PBS before the mechanical strength measurement (tensile test).
2. Insert an adhesive sponge gasket between the load cell and the tested samples.
3. Cut the scaffolds into the dimensions of 1 × 6 cm, and the tensile strengths of collagen samples are measured up to the point when they are broken.
4. Calculate the mechanical parameters of the samples by using a material-testing system at a crosshead speed of 5 mm/min.
5. Verify the extent of porosity by the Archimedes' principle [6].
6. Measure the exterior volume (Vs) of the scaffold (1 × 1 cm) using a Vernier caliper.
7. Soak the scaffold in a pycnometer containing 99% ethanol solution and then calculate the final volume (V_a) of the scaffold through the formula below:

$$V_\alpha = (W_w - W_o) - (W_t - W_p) \times 0.789\ \text{g/cm}^3$$

where W_w is the weight of the ethanol and the pycnometer; W_o is the dry weight of the pycnometer; W_t is the combined weight of the ethanol, the pycnometer, and the scaffold sample; W_p is the combined weight of the dry pycnometer and dry plug sample; and 0.789 g/cm^3 is the density of 99% ethanol solution.

8. Determine the porosity of the collagen scaffold using the formula below:

$$\text{Porosity}\ (\%) = (V_s - V_a)/V_s \times 100\%$$

9. Measure the water content (WC) of the collagen scaffold by swelling the sponge in PBS at room temperature.
10. Blot the wet scaffold using filter paper to remove the water adherent to the surface.
11. Calculate the W_C of the scaffold using the formula below:

$$W_C(\%) = (W_w - W_d)/W_d \times 100\%$$

where W_w and W_d are the weights of the wet and dry sponge, respectively.

3.3 Mouse Subcutaneous Scaffold Implantation to Assess Extrinsic Vascularization

1. Obtain ethical approval for all experimental work involving animals.
2. Soak collagen scaffolds in PBS for 30 min before implantation.
3. Prepare the surgical sites with chlorhexidine 0.5% in 70% ethanol solution.

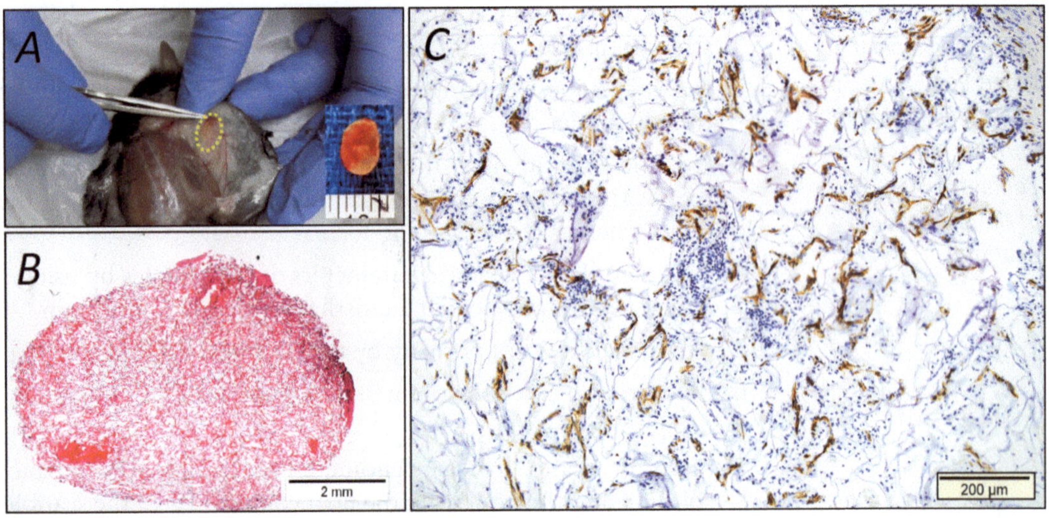

Fig. 2 Vascular infiltration in collagen scaffolds implanted in mice. Appearance of collagen scaffolds (yellow circled) harvested from mice at 2 weeks after implantation (**A**). Representative images of scaffold sections stained with hematoxylin and eosin (**B**) or immunostained with endothelial cell marker, CD31 (brown, **C**)

4. Anesthetize mice with 4% isoflurane inhalation and keep the mice adequately anesthetized throughout the procedure with 2% isoflurane.
5. Make a 0.5 cm midline incision on the dorsal skin with a sterile blade open subcutaneous pockets on either side of the incision using sterile forceps with blunt ends.
6. Insert a collagen scaffold (8 mm in diameter and 2 mm in depth) into each subcutaneous pocket (*see* **Note 1**).
7. Close the incision with sterile stainless steel wound clips.
8. Administer a single dose of Carprofen (5 mg/kg) subcutaneously as analgesic.
9. Euthanize mice at the desired time point of the experiment using a lethal dose of intraperitoneal pentobarbital (350 mg/kg) injection.
10. Harvest the implanted scaffolds for histological analysis (Fig. 2A, B).

3.4 Rat Tissue Engineering Chamber with Flow-Through Femoral Artery and Vein to Assess Intrinsic Vascularization (See Note 2)

1. Obtain ethical approval for all experimental work involving animals.
2. Prepare the surgical sites with chlorhexidine 0.5% in 70% ethanol solution.
3. Anesthetize rats with 4% isoflurane inhalation and keep the rats adequately anesthetized throughout the procedure with 2% isoflurane.

4. Make a 4 cm long skin incision on the groin parallel to the inguinal ligament to expose the femoral vessels.
5. Dissect the epigastric vessels and isolate them from their surrounding fat.
6. Separate the femoral vessels from the surrounding tissues over an approximately 2 cm distance and coagulate all side branches found during dissection with a bipolar coagulator (*see* **Note 3**).
7. Separate intact femoral artery and vein from each other.
8. Impregnate a collagen scaffold with 120 μL of human fibrinogen gel (100 μL of human fibrinogen 15 mg/mL and 20 μL of human thrombin 25 U/mL) and place inside the base of the tissue engineering chamber.
9. Place the tissue engineering chamber base containing a collagen scaffold into the incision area by passing the intact femoral vessels on the corresponding slit of the chamber base. The femoral artery and vein now lie directly over the collagen scaffold (Fig. 3A, B; *see* **Note 4**).

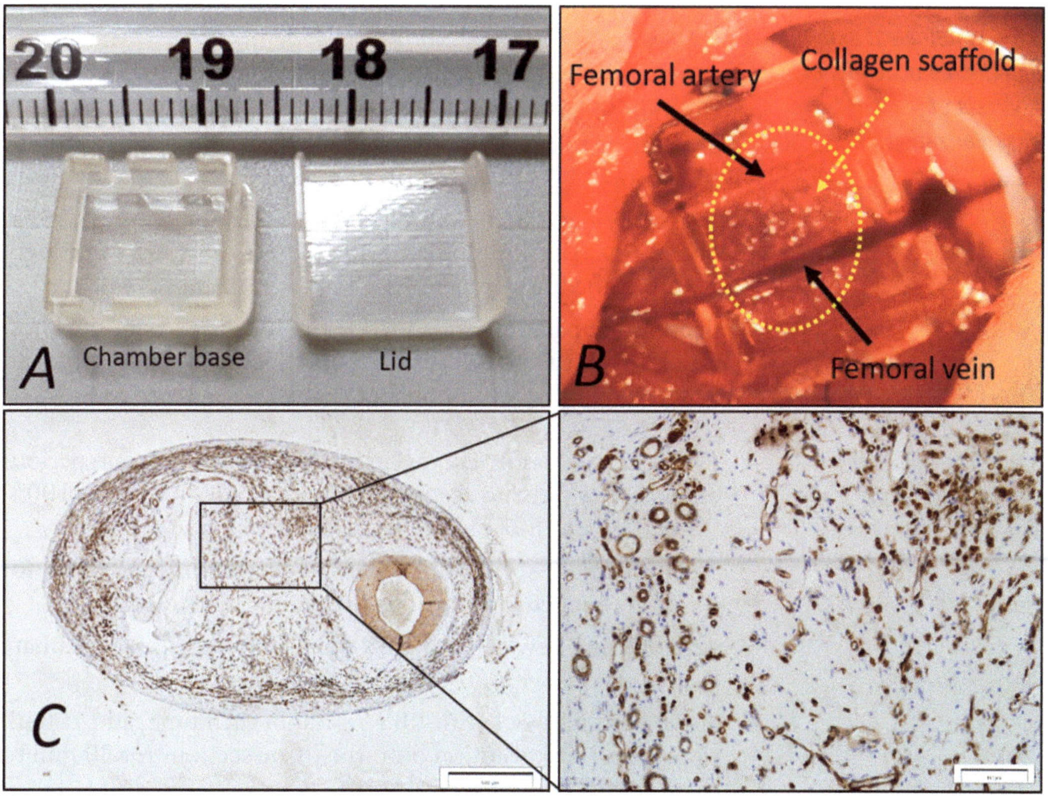

Fig. 3 The vascularized tissue construct generated in rat tissue engineering chambers. The polyacrylic chambers (**A**) containing collagen scaffolds infiltrated with fibrinogen gel (yellow circled) were placed around the femoral artery and vein in the groin region of rats (**B**). Representative images of tissue sections stained with blood vessel marker lectin (brown, **C**). Scale bar: left, 500 μm; right, 100 μm

10. Immobilize the chamber base by suturing it to the surrounding muscle with a 6-0 polypropylene suture.
11. Close the chamber by attaching the lid to the base.
12. Close the wound using continuous running 4-0 silk suture plus two to three additional interrupted stitches.
13. Administer a single dose of Carprofen (5 mg/kg) subcutaneously as analgesic.
14. Once the experiment's time points are reached, anesthetize the rat with isoflurane inhalation.
15. Open the wound and expose the implanted chamber.
16. Ligate the femoral vessels proximal and distal to the chamber and remove the chambers with the containing tissue en bloc.
17. Remove the tissue from the chamber and blot dry the tissues and weigh. Determine the volume of tissue constructs by a volume displacement measurement method [7, 8].
18. At the end of the experiment, euthanize the rat using a lethal dose of intraperitoneal pentobarbital (350 mg/kg) injection.

3.5 Histological and Immunohistochemical Analysis of Vascularization

3.5.1 Tissue Processing

1. Fix tissue in 4% PFA overnight at 4 °C.
2. Rinse the tissue with PBS for 5 min and dehydrate the tissue through serial changes of ethanol (70%, 80%, 95%, and 100%), clear in histolene, and infiltrate with paraffin wax in a tissue processor.
3. For tissue harvested from the rat tissue engineering chamber, divide tissue into multiple transverse sections (1–2 mm thick) before embedding in paraffin.
4. Section the paraffin-embedded tissue blocks at 5 μm thickness and transfer the sections onto glass slides for hematoxylin and eosin staining or immunohistochemistry (*see* **Note 5**).

3.5.2 CD31 and Lectin Immunohistochemical Staining of Blood Vessels

1. Deparaffinize tissue sections in histolene for two 5-min periods and rehydrate tissue sections through graded ethanol (100%, 95%, 70%, and 50%) to tap water for 5 min each.
2. Incubate tissue sections in 3% hydrogen peroxide solution for 5 min to quench endogenous peroxidase activity.
3. After washing twice with PBS for 5 min each time, incubate tissue sections with proteinase K for 8 min.
4. After washing twice with PBS for 5 min each time, add 100 μL of protein block solution onto the tissue sections for 30 min to mask nonspecific antigens.
5. After washing once with PBS for 5 min, incubate tissue sections with 100 μL of rat anti-mouse CD31 (3 μg/mL; for scaffolds harvested from mouse subcutaneous implantation) at room

temperature for 60 min or biotinylated *Griffonia simplicifolia* lectin (6.67 μg/mL, for tissues harvested from rat tissue engineering chambers) at 4 °C overnight.

6. After washing three times with PBS for 5 min each time, incubate tissue with 100 μL of biotinylated rabbit anti-rat IgG (4 μg/mL; for scaffolds harvested from murine subcutaneous implantation) at room temperature for 30 min.
7. After washing three times with PBS for 5 min each time, incubate the sections with 100 μL of avidin-biotinylated-peroxidase complex (Vectastain Elite ABC kit; for scaffolds harvested from mouse subcutaneous implantation) or horseradish peroxidase-streptavidin (1.78 μg/mL; for scaffolds harvested from rat tissue engineering chambers) at room temperature for 30 min. Wash the slides with PBS twice for 5 min each.
8. After washing three times with PBS for 5 min each time, add 100 μL of diaminobenzidine chromogen solution to the tissue sections on the slides to visualize peroxidase activity (*see* **Note 6**).
9. Counterstain the tissue sections with hematoxylin and dehydrate the tissue sections through graded ethanol (70%, 100%, and 100%) to histolene for 5 min each.
10. Coverslip the tissue section in DPX mounting medium and dry overnight before visualization of stained blood vessels (Figs. 2C and 3C).

3.6 Morphometric Analysis of Tissue Vascularization

1. Quantify tissue vascularization by using a video microscopy with computer-assisted stereo investigator system [7].
2. Count three central scaffold cross-sectional sections 200 μm apart (for scaffolds harvested from mouse subcutaneous implantation) or four complete transverse sections (for tissue constructs harvested from rat tissue engineering chambers) with a 20× magnification objective.
3. Use systematic random sampling, superimpose 12-point grids (400 × 400 μm) on randomly selected fields representing 25% of the total area, and record whether each grid point overlies a positive CD31, or lectin positive blood vessel, or overlies other tissue or scaffold material.
4. Determine the percentage of vascular volume by dividing the number of points in each of the selected fields that fell randomly on blood vessels (CD31- or lectin-positive blood vessels including the wall thickness and the lumen) by the total number of points counted for that tissue section and multiply by 100.
5. Calculate the absolute vascular volume of scaffolds by multiplying the percentage of vascular volume by the total tissue volume (*see* **Note 7**).

4 Notes

1. Keep the scaffold away from the macrovascular.
2. Please refer to [5] for video illustration of the procedure.
3. Make sure that the femoral vessels are at rest in their natural position without any twists.
4. Make sure that there are no twists or kinks of the vessels.
5. Allow the slides to dry overnight and store slides at room temperature until ready for use.
6. Diaminobenzidine is carcinogenic. Care should be taken when handling and disposing of diaminobenzidine.
7. All counting should be performed by a trained personnel blinded to the identity of the tissue samples.

Acknowledgments

The authors declare no conflict of interest. This work was supported by grants from The National Health and Medical Research Council of Australia (NHMRC#1061912), Rebecca L Cooper Medical Research Foundation, St Vincent's Hospital (Melbourne) Research Endowment Fund, and Stafford Fox Medical Research Foundation. J.H.W. received a R.B. McComas Research Scholarship in Ophthalmology, Gordon P Castles Scholarship, and a Melbourne Research Scholarship. The Centre for Eye Research Australia and the O'Brien Institute Department of St Vincent's Institute of Medical Research received Operational Infrastructure Support from the Victorian State Government's Department of Innovation, Industry and Regional Development.

References

1. Yannas IV, Tzeranis DS, Harley BA, So PTC (2010) Biologically active collagen-based scaffolds: advances in processing and characterization. Philos Trans A Math Phys Eng Sci 368:2123–2139
2. Abou Neel EA, Bozec L, Knowles JC et al (2013) Collagen—emerging collagen based therapies hit the patient. Adv Drug Deliv Rev 65:429–456
3. Shoulders MD, Raines RT (2009) Collagen structure and stability. Annu Rev Biochem 78:929–958
4. Chan EC, Kuo S-M, Kong AM et al (2016) Three dimensional collagen scaffold promotes intrinsic vascularisation for tissue engineering applications. PLoS One 11:e0149799
5. Zhan W, Marre D, Mitchell GM et al (2016) Tissue engineering by intrinsic vascularization in an in vivo tissue engineering chamber. J Vis Exp. https://doi.org/10.3791/54099
6. Wan Y, Xiao B, Dalai S et al (2009) Development of polycaprolactone/chitosan blend porous scaffolds. J Mater Sci Mater Med 20:719–724
7. Lim SY, Hsiao ST, Lokmic Z et al (2012) Ischemic preconditioning promotes intrinsic vascularization and enhances survival of implanted cells in an in vivo tissue engineering model. Tissue Eng Part A 18:2210–2219
8. Scherle W (1970) A simple method for volumetry of organs in quantitative stereology. Mikroskopie 26:57–60

Chapter 18

Human Reconstructed Skin in a Mouse Model

Jun Mi, Shuai Chen, Lin Xu, Jie Wen, Xin Xu, and Xunwei Wu

Abstract

Currently, no ideal in vivo skin model, to exactly mimic the native human skin, has been utilized for laboratory and clinical application. Here, we describe a method to in vivo reconstitute a human skin model, so-called hRSK, by using culture-expanded skin cells. We grafted a mixture of dissociated human epidermal and dermal cells onto an excision wound on the back of immunodeficient mouse to generate the hRSK, and the hRSK, containing epidermis, dermis, and subcutis and also appendages such as hair follicles, histologically mirrors in situ human skin.

Key words Mouse model, Skin epidermal cells, Skin dermal cells, Human reconstructed skin, Cell grafting

1 Introduction

The reconstituted human skin model is an important tool for studying normal and abnormal biology of skin, including cancer, pigmentation, toxicity, and wound healing [1–4]. Currently, several in vitro human skin-equivalent models are commercially available and theses models, as alternatives to animal testing, have been mainly applied for testing irritation, corrosion, phototoxicity of skin drugs and cosmetics, as well as basic research. However, these skin substitutes, which are made of collagen sheets with embedded fibroblasts alone [5] or with overlaid epidermal cells [6], are not real human skin, because they lack normal skin components and structures, such as subcutis, and appendages [7–10]. The same problem exists for the in vivo model, which is generated by transplantation of tissue engineering skin onto an immunodeficient mouse [11–13]. On the other hand, the transplanted full- or split-thickness human skin mouse model has also been available for skin research, but it is difficult to maintain the normal skin structure after the graft is taken, and importantly the transplanted human skin is hard to be modified, such as genetic modification, upon a research purpose [14, 15]. Therefore, it is

Sophie Böttcher-Haberzeth and Thomas Biedermann (eds.), *Skin Tissue Engineering: Methods and Protocols*, Methods in Molecular Biology, vol. 1993, https://doi.org/10.1007/978-1-4939-9473-1_18,

essential to reconstitute a skin model, which structure should be histologically similar to native human skin, by culture-expanded human cells for laboratory and clinical applications. In recent years, some progresses have been made on reconstitution of a human skin containing hair follicles [16–18], and especially we recently reported a skin model, named human reconstituted skin (hRSK), by cultured human cells, containing epidermis, dermis, and subcutis with regeneration of mature hair follicle and sebaceous glands, which mimics in situ human skin [19]. Moreover, our study showed the healing procedure of the hRSK after wound is similar to that of native human skin [20]. In order to extend this mouse model, here we describe the method in details. The protocol contains the isolation and culture of both epidermal and dermal cells from skin tissues; grafting a cell slurry of mixture of epidermal and dermal cells at a certain ratio onto an immunodeficient mouse; an observation of graft appearance after transplantation; and finally the histological and immunofluorescence analysis of hRSK. The hRSK in a mouse model promises to be valuable as a laboratory model for studying biological, pathological, and pharmaceutical problems of human skin.

2 Materials

2.1 Isolation of Skin Cells

1. Fetal scalp tissues, neonatal and adult human foreskins (the procedure for obtaining human fetal and foreskin tissues from discarded hospital specimens without any personal identity information was approved by the Medical Ethical Committee of the School of Stomatology Shandong University (NO.2015120401, Date: 12-05-2015).
2. 2.5 mg/mL Dispase in phosphate-buffered saline (PBS).
3. 2.5 mg/mL Collagenase in Dulbecco's modified Eagle medium (DMEM).
4. Neutralization solution: 10% Fetal bovine serum (FBS) in DMEM.
5. 0.05 and 0.25% trypsin (porcine).
6. Coating Matrix kit.
7. Rock inhibitor Y-27632: 10 μm in distilled water.
8. Forceps and scalpel blades.
9. 100 μm Filter.

2.2 Cell Culture

1. Dermal cell culture medium: DMEM/F12 (3:1) containing 0.1% antibiotics (penicillin/streptomycin), 40 mg/ml fungizone, 20 ng/mL recombinant human fibroblast growth factor

(hFGF), 20 ng/mL recombinant human epidermal growth factor (hEGF), 2% B27 supplement with 5% FBS.

2. Epidermal cell culture medium: CNT07 (CELLnTEC) plus 0.1% penicillin/streptomycin.

2.3 Grafting

1. Mice: Nu/Nu nude or SCID.
2. Anesthetic: 10 mg/mL Ketamine mixed with 1 mg/mL xylazine in PBS.
3. Surgical tools: Scissors, forceps, tweezers, sutures, needles, medical sterile tape, antimicrobial ointment, Vaseline, 1 mL syringe, polyethylene terephthalate membrane (PET, 3.0 μm pore size).

2.4 Histology and Immunofluorescence

1. OCT (Tissue-Tek).
2. 4% Paraformaldehyde (PFA).
3. 85% Ethanol, 95% ethanol, 100% ethanol.
4. Eosin and hematoxylin.
5. Xylene and permount (xylene based).
6. VECTASHIELD® Mounting Media with DAPI (Vector, H-1200).
7. Blocking buffer: 2% Bovine serum albumin and 5% donkey serum in PBS with 0.01% Triton X-100.
8. Primary and secondary antibodies and dilution times (antibodies were diluted in blocking buffer): Rat anti-FITC-conjugated (CD49) α6-integrin (StemCell; Cat. No. 10111, 1:100); mouse anti-human pan-cytokeratin (BD; Cat. No. 550951, 1:400); monoclonal mouse anti human-vimentin (5G3F10) (Cell Signaling; Cat. No. 3390, 1:200); mouse anti-human nuclei antibody (Millipore; Cat.No.MAB1281, 1:200); Alexa-Fluor-594 donkey anti-mouse IgG (Cat. No. A21203).

3 Methods

3.1 Cell Isolation and Culture

1. The skin tissue (*see* **Note 1**) was labeled and weighed, and then it was transferred into a petri dish.
2. The tissue was rinsed three times in 70% ethanol, and washed with PBS plus 2× antibiotics three times, for 5 min each time. Then the washed skin tissue was transferred into a new petri dish, and was cut into 5 × 5 mm strips with fine scalp, and incubated in dispase solution overnight at 4 °C (*see* **Note 2**). The next day, the epidermis and dermis were mechanically separated by using forceps.
3. For dermal cell preparation: The dermis was minced using crossed scalpel blades (*see* **Note 3**) and incubated in collagenase

at 37 °C water bath with shaking; 10 mL collagenase solution was added for 1 g tissue.

After 30-min incubation, the dermal solution was neutralized in DMEM with 10% FBS, and passed through 100 μm filter, and the filtrate centrifuged and rinsed with PBS.

The pellet of dermal cells was plated in dermal cell culture medium; medium was changed every 5 days. Dermal cells were passaged when the cell density attained 80–90% confluence; the passage dilution was 1:5–10.

4. For epidermal cell preparation: The peeled epidermis was transferred to a new petri dish, and was homogenized by scissors. The homogenized epidermis was incubated in 0.05% trypsin at 37 °C water bath with shaking for 15 min, 5 mL Trypsin solution was used for around 1 cm^2 size of skin tissue. The incubation solution was neutralized with 10% FBS in DMEM, filtered, centrifuged, and washed with 10% FBS DMEM once.

   The dissociated epidermal cells were plated into culture vessels, which were precoated with coating matrix, by resuspending with CNT07 plus 10 μm Rock inhibitor Y-27632. After 3-day culture, Y27632 was removed; medium was changed every other day.

   Epidermal cells were passaged when the cell density attained 80–90% confluence, and the passage dilution was 1:3. Both dermal and epidermal cells were collected for grafting after the density reached 90–100% confluence. To allow adequate expansion, all cells used for grafting were passaged from one to four times. The extra cells can be frozen in liquid nitrogen for the next-time use.

3.2 Grafting Dissociated Human Cells onto Skin Wounds of Immunodeficient Mice

1. 2×10^6 Dissociated cultured epidermal cells mixed with 3×10^6 dermal cells (total five million cells) were suspended a total volume of 100 μL of F12 as a cell slurry (*see* **Note 4**).
2. The cell slurry was then transferred onto a PET membrane of 6-well cell culture insert (*see* **Note 5**) and incubated at 37 °C for 1–1.5 h (*see* Fig. 1a), and it is ready for grafting (*see* **Note 6**).
3. The mouse was anesthetized by I.P. injection of a mixture of ketamine (100 mg/kg body weight) and xylazine (10 mg/kg body weight). The back skin of mouse was wiped with 70% ethanol and the two graft-receiving sites on the back of mouse were marked before excision (*see* Fig. 1b). Each mouse could be received one or two grafts. The size of the excision wound was determined by the size of the cell slurry.
4. The host full-thickness skin was excised, and the membrane with the cell slurry was placed onto the wound (cell side faced down, *see* **Note 7**) and sutured into the host skin (*see* Fig. 1c).
5. After the membrane was sutured to the host skin, the grafting site was covered with antimicrobial ointment and Vaseline and then wrapped by sterile tape (*see* Fig. 1d).

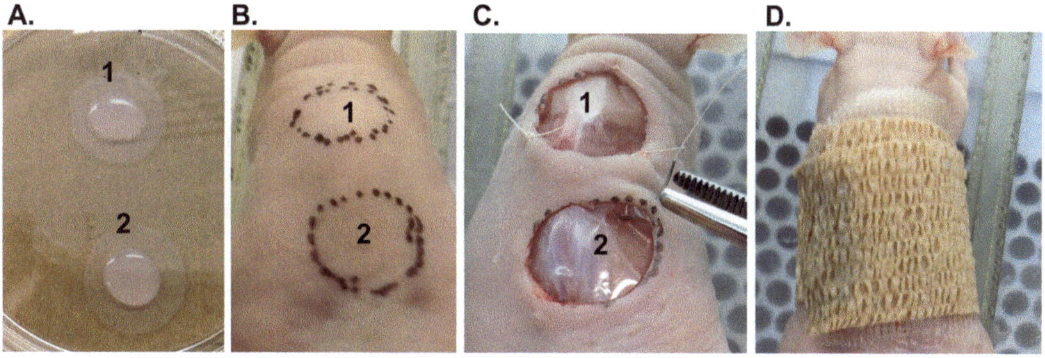

Fig. 1 A typical grafting procedure. (**a**) The two cell slurries labeled as 1 and 2 were placed onto the PET membrane individually. (**b**) The two graft-receiving sites on the back of mouse were marked before excision. (**c**) The host full-thickness skin was excised, and the membrane with the cell slurry was placed onto the wound (cell side faced down) and sutured into the host skin. (**d**) The grafted mouse was wrapped with bandage as a final procedure

6. The dressings were removed after 1 week and the PET membrane was spontaneously shed as the outer epidermis formed (*see* **Note 8**).
7. The grafted mice were monitored weekly, and the photograph of graft was taken regularly (*see* Fig. 2). The pigmented hRSK was usually distinguished from the host skin at around 1 month (Fig. 2a), and with formation of hair follicles the hRSK grew thicker which was above the host skin at around 2 months (Fig. 2b). At 3 months, the reconstituted and pigmented hairs could reach around 1 cm long (Fig. 2c), and the length of hair could reach around 3 cm at 6 months after grafting (Fig. 2d).

3.3 Histological Analysis (HE Stain) of Human Reconstituted Skin

1. The histological structure of hRSK was analyzed by hematoxylin/eosin (HE) stain. The hRSKs were collected at desired time points. The skin was fixed in 4% PFA and embedded in optimal cutting temperature compound (OCT) for cryosection. 10 μm Cryosections were taken for histological analysis following the standard protocol. Briefly, the section was incubated with hematoxylin for 2 min, and washed with distilled H_2O; the sections were incubated with eosin for 30 s followed by the dehydration procedure, which were treated with a series of ethanol solution of increasing concentration until 100% alcohol is reached. After dehydration, the sections were washed with xylene three times, and mounted with Permount (xylene-based solution). The stained sections were analyzed by microscope.
2. The HE stain revealed that the hRSK formed as early as 1 week after grafting [19]. At 3–4 weeks, the hRSK was a well-organized structure, and the morphogenesis of hair follicles

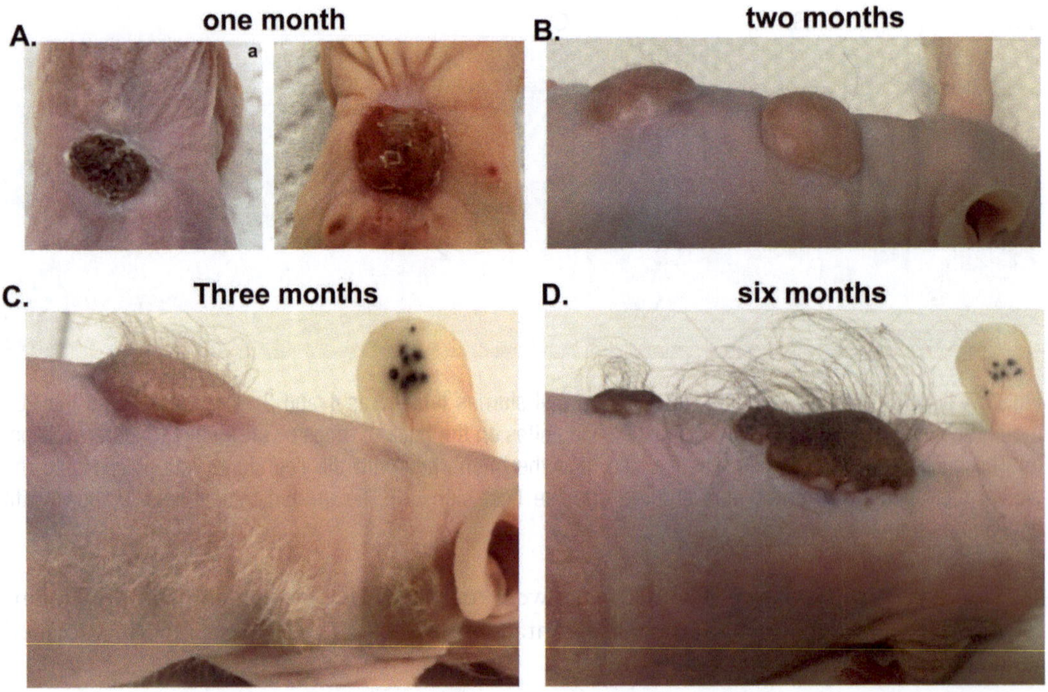

Fig. 2 The images of hRSK at different time points after grafting. (**a**) The image of reconstituted skin from two mice grafted with different donor cells at 1 month after grafting. (**b**) The image of two reconstituted skins from the same cells at 2 months after grafting, and hair follicle formation can be observed. (**c**) The images of 3-month reconstituted skin clearly showed hair growing. (**d**) The image of 6-month hRSK with much longer hairs, and the hRSK is pigmented

was initiated at this time with formation of placode and dermal condensate (Fig. 3a) (*see* note, and the stage 8 hair follicle was formed around 8 weeks) (Fig. 3b). At 12 weeks, anagen phase of mature hair follicles was observed to be associated with sebaceous gland and dermal papilla (Fig. 3c). At 6 months, the structure of hRSK was similar to that of nature human scalp tissue (Fig. 3d).

3.4 Immunofluorescence Staining of Human Reconstituted Skin

1. To confirm that the hRSK was formed from human cells, immunofluorescence analysis of human specific antibodies was performed as follows: the cryosection was fixed in 4% paraformaldehyde, incubated with the blocking buffer at room temperature for 1 h, and then applied with primary antibody at 4 °C overnight. On the second day, after removing the primary antibody solution, the specimen was incubated with secondary antibody at room temperature. After 1-h incubation, the section was washed with PBS and mounted mounting medium with DAPI, and the staining was analyzed by immunofluorescence microscope.

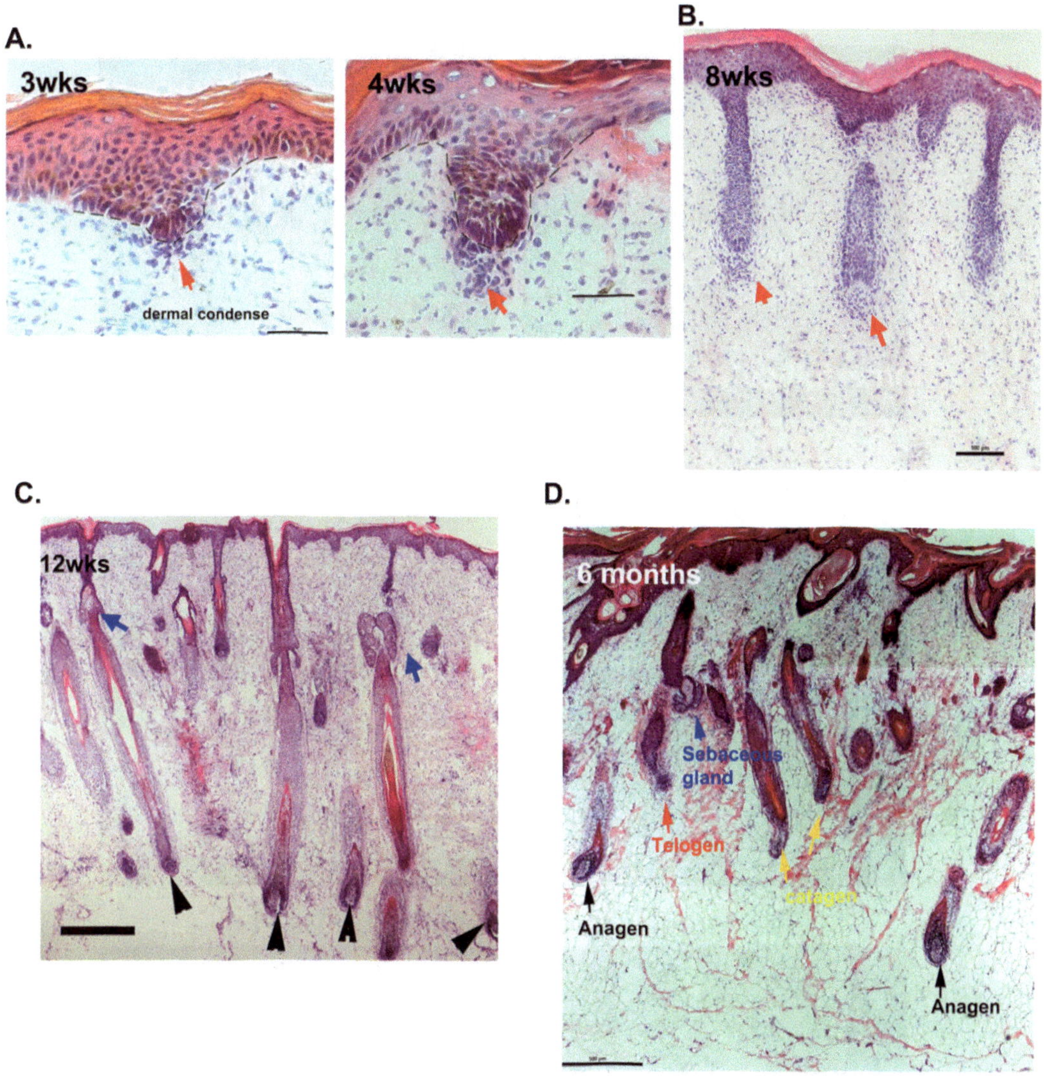

Fig. 3 Histological analysis (HE staining) of hRSK at different time points. (**a**) The 3-week- and 4-week-old hRSK were harvested and showed formation of well-organized skin structure with early morphogenesis of hair follicle formation (placode formation around 3 weeks), and red arrows indicate dermal condense, early stage of dermal papilla. Bars = 50 μm (**b**). HE staining of 8 weeks hRSK showed a well-differentiated and -organized skin with formation of nearly mature hair follicles. Red arrows point dermal papilla. Bar = 100 μm (**c**). At 12 weeks, the skin grew thicker, and mature anagen hair follicles associated with sebaceous glands (blue arrows) and dermal papilla (black arrows) were observed. Bar = 100 μm (**d**). The histological structure of 6-month hRSK was similar to a normal human scalp tissue, which contains epidermal layer, dermal layer, and subcutaneous layer (subcutis), and contains different stages of hair follicles associated with sebaceous glands and dermal papilla. Bar = 500 μm

2. Figure 4 shows that 6-week-old hRSKs were positive for all human specific antibodies which include human pan-ck antibody for epidermal cells, human vimentin antibody for human dermal cells, and human nuclei antibody for all human cells.

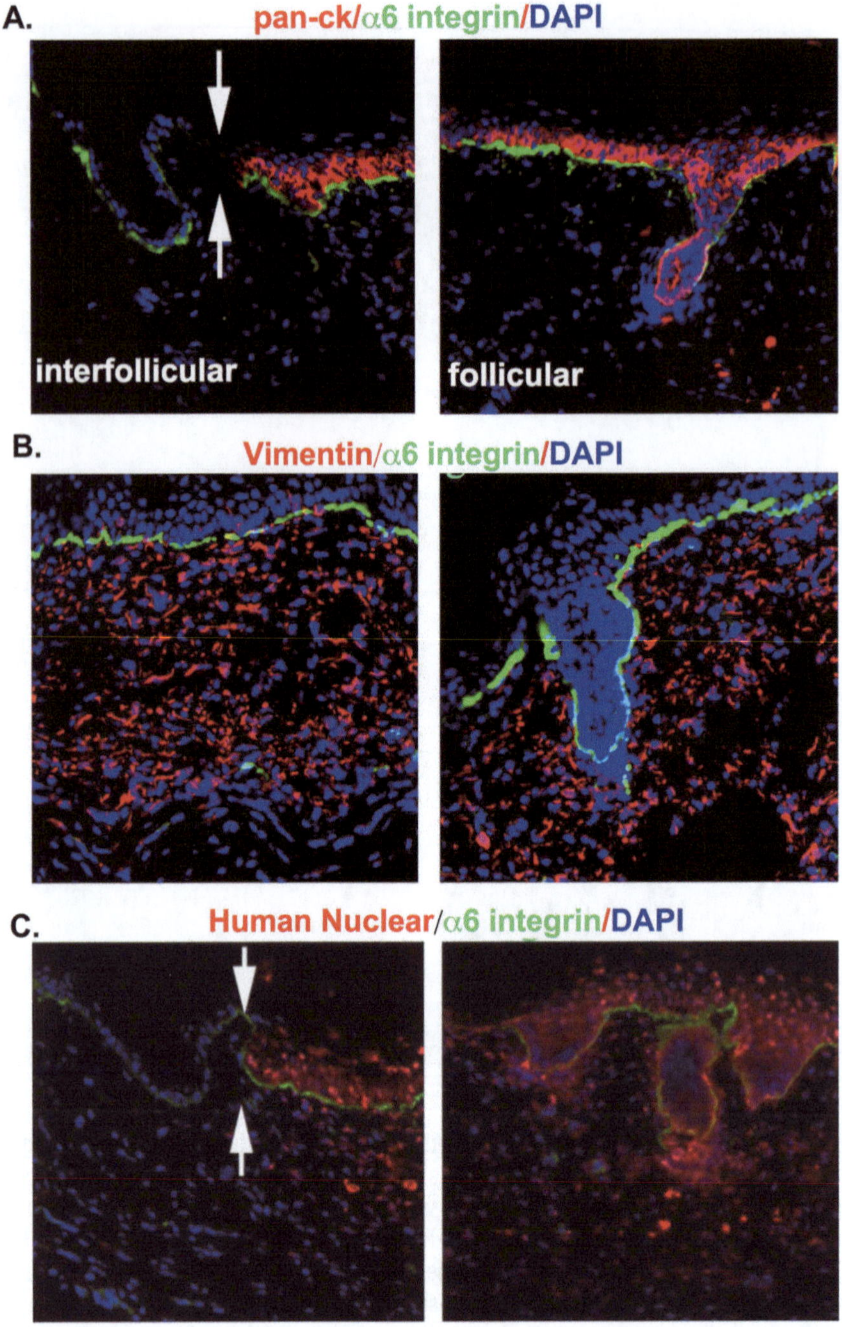

Fig. 4 The hRSK is made of human cells. (**a**). Human pan-cytokeratin (pan-ck) was present in the epidermis and early follicles, and was absent in the mouse tissue; the white arrows indicate the margin of human and mouse skin. (**b**) The dermal cells of hRSK were positive for human vimentin antibody. (**c**) Both epidermal and derma cells of hRSK were positive for human nuclei antigen, but the mouse tissue was negative; the white arrows indicate the border of host mouse skin and hRSK. α6-Integrin (green), a basal membrane component, stains the junction of epidermis and dermis of hRSK, and DAPI stains cell nuclei (blue)

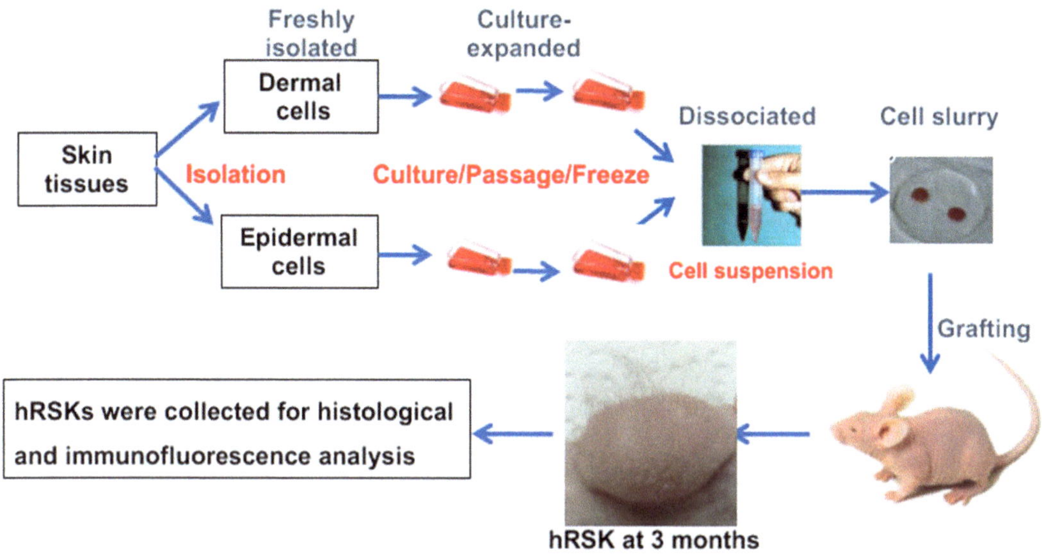

Fig. 5 The scheme to reconstitute human skin in a mouse model

Importantly, the mouse tissues were negative for these human antibodies.

3.5 The Whole Experimental Procedure Was Summarized in Fig. 5

1. Figure 5 shows all steps involved to in vivo reconstitute a full-thickness hRSK. The isolated skin cells can be passaged and frozen down if needed (*see* **Note 9**).

4 Notes

1. The skin tissue, collected from clinics, should be avoided for strong shaking during the transportation, and it should be always kept in cold DMEM or PBS before using for isolating cells. We tested that the skin tissue can be stored in fridge for up to 72 h without significantly affecting isolated cell viability.
2. In order to get efficient separation of epidermis from dermis, the skin tissue should be trimmed for taking fat tissue and blood clot away.
3. In order to get a high yield of dermal cell preparation, the well-homogenization of dermis is crucial. Usually, the homogenizing procedure of 1 g tissue takes 10–15 min.
4. Generally, we used total five million cells, which mixed two million epidermal cells and three million dermal cells, for generating one graft, which covered around 1 cm diameter excision wound. We tested what is the optimal ratio between epidermal and dermal cells for better efficiency of hair

formation in hRSK, and we found that the ratio of 1:1.5–2 was the best ratio to efficiently reconstitute hair follicles.

5. A cell slurry is actually a big aggregate formed from the mixture of epidermal and dermal cells, and the aggregate should attach the PET well. If cells cannot stick each other cohesively and failed to form aggregate, it indicates the quality of cells not optimized, for example, the viability of cell was low, or the cells were differentiated.
6. After 1–1.5-h incubation, the cell slurry could be stored for up to 6 h at 4 °C, but it is better to graft the cells as soon as possible.
7. Before placing the membrane with the cell slurry onto the wound, the extra medium should be removed by gently lifting the membrane with forceps to touch a sterile tissue towel to absorb the extra medium.
8. The PET membrane did not need to be removed since it was easy to be broken and lost once the wrapping bandage was taken away.
9. In order to generate a hair-producing hRSK, the dermal cells must be derived from a scalp tissue, but the epidermal cells could be from any site of skin. The hRSK examples from Fig. 2 to Fig. 4 were reconstituted by the mixture of fetal scalp dermal cells and neonatal foreskin epidermal cells. The number of passages significantly affects the efficiency of hair formation in the hRSK, and we found that the expanded cells were difficult to reconstitute hair follicles after passage 5. The early-passage cells could maintain their regeneration patency after freezing down; therefore the culture-expanded cells can be stored in liquid nitrogen for several years; however it is not suggested to keep cells in −80 °C freezer for more than 2 weeks.

Acknowledgments

This work was supported by National Key Research and Development Program of China (2017YFA0104604), General Program of National Natural Science Foundation of China (81772093), and Shandong Taishan Scholar Award (tshw201502065).

References

1. Reuter C, Walles H, Groeber F (2017) Preparation of a three-dimensional full thickness skin equivalent. Methods Mol Biol 1612:191–198
2. Desmet E, Ramadhas A, Lambert J, Van Gele M (2017) In vitro psoriasis models with focus on reconstructed skin models as promising tools in psoriasis research. Exp Biol Med (Maywood) 242(11):1158–1169
3. Vinardell MP, Llanas H, Marics L, Mitjans M (2017) In vitro comparative skin irritation induced by nano and non-nano zinc oxide. Nanomaterials (Basel) 7(3):56

4. Wills JW, Hondow N, Thomas AD, Chapman KE, Fish D, Maffeis TG, Doak SH (2016) Genetic toxicity assessment of engineered nanoparticles using a 3D in vitro skin model (EpiDerm). Part Fibre Toxicol 13(1):50
5. Naughton GK, Mansbridge JN (1999) Human-based tissue-engineered implants for plastic and reconstructive surgery. Clin Plast Surg 26(4):579–586
6. Parenteau N (1999) Skin: the first tissue-engineered products. Sci Am 280(4):83–84
7. Michel M, L'Heureux N, Pouliot R, Xu W, Auger FA, Germain L (1999) Characterization of a new tissue-engineered human skin equivalent with hair. In Vitro Cell Dev Biol Anim 35 (6):318–326
8. Havlickova B, Biro T, Mescalchin A, Arenberger P, Paus R (2004) Towards optimization of an organotypic assay system that imitates human hair follicle-like epithelial-mesenchymal interactions. Br J Dermatol 151(4):753–765
9. Larouche D, Cuffley K, Paquet C, Germain L (2011) Tissue-engineered skin preserving the potential of epithelial cells to differentiate into hair after grafting. Tissue Eng Part A 17 (5–6):819–830
10. Krugluger W, Rohrbacher W, Laciak K, Moser K, Moser C, Hugeneck J (2005) Reorganization of hair follicles in human skin organ culture induced by cultured human follicle-derived cells. Exp Dermatol 14(8):580–585
11. Geer DJ, Swartz DD, Andreadis ST (2004) In vivo model of wound healing based on transplanted tissue-engineered skin. Tissue Eng 10 (7–8):1006–1017
12. Escamez MJ, Garcia M, Larcher F, Meana A, Munoz E, Jorcano JL, Del Rio M (2004) An in vivo model of wound healing in genetically modified skin-humanized mice. J Invest Dermatol 123(6):1182–1191
13. Martinez-Santamaria L, Guerrero-Aspizua S, Del Rio M (2012) Skin bioengineering: preclinical and clinical applications. Actas Dermosifiliogr 103(1):5–11
14. Maldonado AA, Cristobal L, Martin-Lopez J, Mallen M, Garcia-Honduvilla N, Bujan J (2014) A novel model of human skin pressure ulcers in mice. PLoS One 9(10):e109003
15. Momtazi M, Kwan P, Ding J, Anderson CC, Honardoust D, Goekjian S, Tredget EE (2013) A nude mouse model of hypertrophic scar shows morphologic and histologic characteristics of human hypertrophic scar. Wound Repair Regen 21(1):77–87
16. Li S, Thangapazham RL, Wang JA, Rajesh S, Kao TC, Sperling L, Darling TN (2011) Human TSC2-null fibroblast-like cells induce hair follicle neogenesis and hamartoma morphogenesis. Nat Commun 2:235
17. Thangapazham RL, Klover P, Wang JA, Zheng Y, Devine A, Li S, Darling TN (2014) Dissociated human dermal papilla cells induce hair follicle neogenesis in grafted dermal-epidermal composites. J Invest Dermatol 134 (2):538–540
18. Higgins CA, Chen JC, Cerise JE, Jahoda CA, Christiano AM (2013) Microenvironmental reprogramming by three-dimensional culture enables dermal papilla cells to induce de novo human hair-follicle growth. Proc Natl Acad Sci U S A 110(49):19679–19688
19. Wu X, Scott L Jr, Washenik K, Stenn K (2014) Full-thickness skin with mature hair follicles generated from tissue culture expanded human cells. Tissue Eng Part A 20 (23–24):3314–3321
20. Wen J, Li X, Leng X, Xu X, Wu X (2017) An advanced mouse model for human skin wound healing. Exp Dermatol 26(5):433–435

Chapter 19

Pig Model to Test Tissue-Engineered Skin

Christian Tapking, Daniel Popp, and Ludwik K. Branski

Abstract

Tissue engineering of skin is a field with high research activities and major importance for wound healing, especially following burn injuries. Animal models enable to test tissue-engineered skin as well as different types of cells in a realistic setting. Although there are several challenges in working with pigs, it is a good model because of its similarity to the human skin and a comparable wound regeneration. Here, we explain our routinely used methods for using pig models to test tissue-engineered skin in burn injuries.

Key words Tissue engineering, Skin graft, Pig model, Burn injury, Wound healing

1 Introduction

There is a high need for effective and safe wound coverage in different clinical settings such as genetic disorders (e.g., bullous conditions), acute trauma, and chronic wounds. One of the major reasons for the need of tissue repair is thermal trauma since the possibility of skin regeneration in this type of injury is often very limited. This may lead to deep wounds and massive (hypertrophic) scarring, which can affect the patients in terms of a decreased quality of life or delayed reintegration into society during their lifetime [1–3]. Wound healing is of major importance in the clinical outcome and survival of burn patients. Early excision and grafting reduce metabolic responses and increased survival after burn injury [4–6]. However, early excision leaves large uncovered wounds that need to be covered with either the patient's own non-burned skin, cadaver skin, or other synthetic/biologic material. Because of the rather limited access to autografts and allografts, bioengineered skin became an active field in research and a great promise, especially in the treatment of burns.

Clearly, human studies are the most accurate way of investigating the effectiveness of wound treatment, but are often impractical because of the limited histological assessment, which requires

Sophie Böttcher-Haberzeth and Thomas Biedermann (eds.), *Skin Tissue Engineering: Methods and Protocols*, Methods in Molecular Biology, vol. 1993, https://doi.org/10.1007/978-1-4939-9473-1_19,

frequent biopsies. Furthermore, ethical concerns prevent the use in a regular basis [7]. There are numerous in vitro and in vivo wound-healing models. Many models can only isolate specific mechanisms or events and not the impact on the overall human organism. Studies suggest that cultured skin cells returned to patients may keep the competence to self-renew for a lifetime [8]. Currently available bioengineered skin grafts still deal with problems such as patient's safety and clinical efficacy and sheets with keratinocytes are often complicated to handle [2]. The selection of an animal model depends on factors such as availability, anatomical and physiological similarity, and costs. Small animals such as rat, rabbit or mouse are frequently used, inexpensive, and good to handle. However, the small animals differ significantly from humans regarding the anatomical and physiological structure of their skin [9, 10]. Using pigs as a model is associated with high costs and the need of a good infrastructure for the handling of large animals. Nevertheless, pig is a suitable model because of its anatomic and physiological similarity to human skin in many terms [9]. Epidermis thickness in pigs ranges from 30 to 140 μm in pigs and from 50 to 120 μm in humans [11]. In addition, both show similar dermal collagen, papillary bodies, and abundant subdermal adipose tissue and hypertrophic scarring can be assessed appropriately in pigs [12, 13]. Furthermore, the skin of other animals such as rats primarily heals by contraction, whereas the skin of Yorkshire and Red Duroc pigs heals by re-epithelialization, just as seen in the human wound. Furthermore, pigs are better for studies on wound healing due to their size and the instruments, which are used in the daily clinical setting.

In this chapter, we describe our routinely used methodology for pig models to test tissue-engineered skin using the example of burn injuries.

2 Materials

2.1 Animals

All preparations and procedures are done at room temperature if not indicated otherwise.

For the experiments, we usually use Yorkshire or Red Duroc pigs. Housing and animal care are supplied in line with the National Research Council guidelines [14].

1. The healthy 10–40 kg (1–4 months old) pigs are housed individually for at least 1 week for acclimatization at the large animal resource center before starting the surgical procedures.
2. For the duration of the experiment, the animals are held in individual cages with about 3 m^2 size. Cages are cleaned at least once daily.

3. A standard ration of porcine diet (LabDiet® 5084, PMI Nutrition, IN, USA) is given twice daily and water ad libitum.
4. Before undergoing any anesthesiological or surgical procedure (e.g., burning, dressing changes, biopsies), the animals are fasted overnight to achieve safe airway conditions during the anesthesia.

2.2 Surgical Materials and Veterinary Drugs

1. TKX solution: 1 mL TKX contains 0.5 mg ketamine, 0.5 mg xylazine, and 1 mg telamine/zolazepam.
2. Bear Hugger® blanket (Arizant Healthcare Inc., Eden Prairie, MN, USA).
3. Cefazolin (Cefazol® Bosch Pharmaceuticals, IN, USA).
4. 500 mg Levofloxacin tablets (Levaquin®, Ortho-McNeil Pharmaceuticals, TX, USA).
5. Transdermal fentanyl patches,
6. Meeker gas burner.
7. Ringer's lactate solution.
8. 0.1% w/v Trypsin, and 1% trypsin.
9. 5 mM EDTA.
10. 0.5% Collagenase A.
11. 1% Nystin.
12. 2% Polymyxin B/Bacitracin.

3 Methods

In our institution, all animal experiments are conducted in a central animal research facility that provides all required and needed materials. All animals are monitored continuously by professional staff in order to correct postoperative problems such as dressing failures, pain, or bleedings early and effectively.

3.1 Anesthesia and Perioperative Care

Appropriate methods for sedation, inducing anesthesia and pain control, are of major relevance throughout the whole experiments and require professional and experienced staff and guidelines. Animals were fasted overnight to avoid vomiting and aspiration.

1. Anesthesia is inducted by an intramuscular injection of 0.04 mL/kg TKX (*see* **Note 1**).
2. The animals are weighted and clipped before beginning any surgical procedure (*see* **Note 2**).
3. After transfer to the operating room, the pigs are placed on the operation table in prone position, pre-oxygenated with 100% oxygen via inhalation mask, and intubated with a straight

laryngoscope blade using endotracheal tube of size 7 or 8 depending on the pigs' size.

4. Isoflurane is delivered after successful intubation as oxygen: nitrous oxide mixture (2:1) to maintain anesthesia (Aestive® 5, Draeger Medical Inc., PA, USA, *see* **Note 3**).
5. Before beginning with any surgical procedure, animals are depilated using a disposable razor blade.
6. A venous catheter in the right jugular vein, tunneled underneath the skin of the lateral neck and released through the dorsal neck region, was placed before the initial burn (*see* **Note 4**).
7. During the surgical procedure, heart rate and oxygen saturation are monitored via ear oximeter whereas temperature is controlled via rectal thermometer. A disposable Bear Hugger® blanket (Arizant Healthcare Inc., Eden Prairie, MN, USA) is used to maintain a body core temperature of 40 °C.
8. Pigs are allowed to recover on the operating table after each procedure. Once spontaneous breathing is stabilized and gag reflex is apparent, they need to be extubated and transferred to their individual cage. Temperature control is continued in the cage using a heat lamp or blankets. Continuous clinical controls are required by the staff.
9. Perioperative antimicrobial prophylaxis is provided with 1 g Cefazolin i.v. (Cefazol® Bosch Pharmaceuticals, IN, USA) before each operation and dressing and 500 mg Levofloxacin tablets p.o. (Levaquin®, Ortho-McNeil Pharmaceuticals, TX, USA) for 7 days after burn.
10. Postoperative analgesia is provided with transdermal fentanyl patches (Duragesic® 25 or 100, Janssen Pharmaceutics, NJ, USA) applied on the dorsal neck behind the ear and kept in place using surgical staples and an additional protective layer of adhesive transparent dressing (*see* **Note 5**).
11. After surgical procedures, the food intake remains stable without aspiring weight loss.

3.2 Burn Injury

1. Full-thickness contact burns are applied paravertebrally under aseptic conditions by using a heated aluminum bar on the dorsum of the animal. The pigs receive up to 16 bilateral contact burn on the dorsum of 3×3 cm^2. A temperature of 200 °C of the aluminum bar is achieved with a Meeker gas burner and monitored with a digital thermometer (*see* **Note 6**).
2. The heated bar has contact to the animal for 30 s and application pressure is measured with a 50 mL syringe attached to the aluminum bar (*see* **Note 7**).

3. Depending on the study protocol, biopsies from the burned tissue are obtained several times after burn to assess the depth of the burn under experimental conditions (*see* **Note 8**).
4. Throughout and right after the burn procedure, the pigs are provided with Ringer's lactate solution at a rate of 10 mL/kg/h, usually at a total intravenous resuscitation volume of 20 mL/kg. After that, they have free access to water.
5. To mimic the clinical setting, the animals undergo full-thickness excision and grafting 24–48 h after burn. Excision is done in a size of 7×7 cm^2 with the burn area being in the center of the wound.
6. As a supportive procedure after the excision, the area surrounding the wounds can be marked by tattoos to track the rate of wound coverage and re-epithelialization. This can simplify the assessments of scar in the further course of the experiment.

3.3 Producing and Applying the Tissue-Engineered Skin

Producing tissue-engineered skin is currently under intense research and a high amount of different methods are described in the literature. Here, we describe a method that was used at our institution, but can differ from the usually used method at other institutions.

1. Small samples from porcine skin (2×2 cm) are taken, cut into pieces of <0.5 cm^2, and then incubated at 4 °C overnight in 0.1% w/v trypsin.
2. After incubating the epidermal and dermal layers are separated carefully using an anatomic forceps (*see* **Note 9**).
3. The epidermal cells are isolated by incubating at 37 °C in 1% trypsin and 5 mM EDTA for 3 min and centrifuged at $300 \times g$ for 5 min.
4. Then the cells are resuspended into a keratinocyte-cultured medium (*see* **Note 10**).
5. The dermal cells are incubated at 37 °C in a 0.5% collagenase A solution overnight and centrifuged at $300 \times g$ for 10 min the next day. Then the cells are resuspended in fibroblast culture medium again overnight.
6. Dead cells are then washed out using phosphate-buffered saline.
7. Fibroblasts are seeded on a reticular surface of de-epidermized dermis at a concentration of 1×10^5 cells (*see* **Note 11**).
8. The gels with fibroblast are grown in a fibroblast growth medium for 5–7 days and then 1×10^6 keratinocytes are seeded on the surface.

9. After that, the tissue-engineered skin is exposed to air-liquid interface for another 10–14 days with medium changes every 2–3 days.
10. The developed and stable tissue-engineered skin plates are applied on the burned area using surgical staples for fixation.
11. Alternatively to tissue-engineered skin, 1:4 autograft split-thickness mesh grafts can be used as transfer medium for adipose-derived stem cells and other cells.

3.4 Treatment of Donor Site and Wound Coverage

1. Prior to wound coverage, donor sites are labeled in chronologic order from cranial to caudal with ink.
2. Skin biopsies are taken (two central and two peripheral) from each wound site and donor site. If autografts are used to cover the wounds, skin graft donor sites (considered as partial thickness dermal wounds) are covered with wound dressings.
3. The first dressing layer, directly placed on the autograft, consists of non-sticking (silicone) oil emulsion dressing. This layer is covered with a thin protective layer of antimicrobial ointment consisting of 1% Nystin and 2% Polymyxin B/Bacitracin in a Vaseline formulation (*see* **Note 12**).
4. The second layer of double-polyurethane film, such as Opsite® (Smith & Nephew Inc., FL, USA), is placed over the first dressing layer to seal off the wound fields while providing vapor permeability (*see* **Note 13**).
5. A nylon dressing is used as an outer layer for protection of the inner dressing layers.
6. After treatment, the animals should be seen twice daily within the first 2 weeks and daily after 2 weeks. Dressings are checked for being clean and intact to prevent unintended damage to wound dressings caused by normal animal movements (*see* **Note 14**). An adequate and systematic pain assessment as well as frequent check of wound dressing and wounds are essential for the animals' well-being for the outcome of the study.
7. The frequency of dressing changes depends on the specific study protocol (*see* **Note 15**) and is a complex balance between giving the grafts the time to heal and avoiding infections under the dressings. At days 2 through 14 (around days 2, 5, and 14), dressings are changed and skin biopsies (two central and two peripheral for each donor site). Before every dressing change and/or biopsy, the animals' weight should be assessed.
8. When donor sites are healed as well as at the time of scarification (anticipated between 3 and 6 months postinjury), biopsies are taken again (*see* **Note 16**).

3.5 Outcome Measurements

There are several methods to measure the outcome of wound dressings, skin substitutes, or tissue-engineered skin in animal models. Some of them are similar to the methods in human patients because of their noninvasiveness. Others, which for example require frequent biopsies, should only be performed in an experimental setting with animal models.

3.5.1 Clinical Evaluation

Most of the studies concerning tissue-engineered skin use at least one clinical evaluation tool. A typical score as described by Branski et al. is shown in Table 1 [15]. These scores should be assessed by at least two investigators during the dressing changes.

3.5.2 Planimetry

After initial grafting and during each dressing change, we usually take standardized photographs. We use a planimetrical software (Lucia G®, Version 4.8, Laboratory Imaging Ltd., The Czech Republic). Total wound size and size of all open mesh interstices are measured for each wound at each time point (*see* **Note 17**). To assess the healing, total wound size reduction and open interstices have to be compared to the baseline values (date of injury/graft).

3.5.3 Wound Histopathology

After embedded in paraffin and sectioned, histochemical staining techniques (for example hematoxylin-eosin [H&E] staining) can be used to evaluate cellularity, cell morphology, and architectural aspects such as thickness and orientation of the collagen [16]. Differences of the architecture of the collagen, which are very specific for the differentiation between normal skin and scar tissue, are reliably apparent on H&E [17].

3.5.4 Hematology

Selected serum cytokines and hematologic parameters can be used to evaluate the general condition of the animal after surgical procedures and to evaluate systematic effects of the burn or other injury. To later compare the findings with baseline data, blood samples should be obtained prior to the procedure and before grafting. We usually obtain blood samples on days 2, 5, 9, and 14 after the procedure. This can differ in the specific study protocol and should always be discussed before beginning the study. We use a veterinary hematology analyzer (Drew-Scientific Hemavet 950, Escalon Medical Corp, PA, USA) to measure parameters such as WBC, hemoglobin, and platelets. IL-6 and TNF-α are measured by ELISA using porcine kits (P6000 Porcine IL-6 Quantikine Elisa Kit and PTA00 Porcine TNF-α Quantikine Elisa Kit, R&D Systems, Minneapolis, MN, USA). The normal ranges for 20–25 kg Yorkshire/Red Duroc pigs are described in "Swine in the Laboratory" by Swindle and Smith [18]. Some parameters of major interest are shown in Table 2.

Table 1
Clinical evaluation of wounds

Grade	Graft dislocation	Graft adherence	Granulation tissue	Hypergranulation	Hematoma	Fibrin deposition
0	None	Full graft loss	None, full depth defect	None	None	None
1	5–10%, tissue still fully viable	Graft 20% adherent, 50% graft loss	Low level of granulation tissue	10%	Up to 10% wound size	Small patches, up to 10% wound size
2	10–25%, one or two spots, q0% tissue not viable	Graft 50% adherent, 30% graft loss	Granulation tissue level at half of initial wound depth	20%	Up to 20% wound size	Thicker, confluent patches, up to 20% wound size
3	25–50%, more than 20% tissue not viable	Graft 75% adherent, 20% graft loss	Granulation tissue level just below surrounding tissue	40%	Up to 40% wound size	Up to 40% wound size
4	50–80%, more than 50% tissue not viable	Graft 90% adherent	Granulation tissue at level with surrounding tissue	65%	Up to 60% wound size, partial graft loss	Up to 60% wound size
5	Graft completely dislocated, full tissue loss	Full graft adherence	Fully granulated wound, exceeds level of surrounding tissue	100%	Up to 100%, total graft loss	Wound fully covered with fibrin and graft skin totally macerated

Modified from Branski et al. [11]

Table 2
Normal ranges of blood parameters in pigs

Measurement	Mean	Range
Blood characteristics		
Hematocrit (%)	27.0	23.5–33.0
Hemoglobin (g/dL)	8.5	7.3–10.2
WBC/μL	4500	3800–10,000
Platelet count (10^9/L)	300	250–600
Plasma albumin (g/L)	25.4	21.6–28.0
Sodium (mEq/L)	138	129–143
Potassium (mEq/L)	4.4	3.9–4.1
Blood gas and acid-base status		
Arterial PO_2 (mmHg)	82	73–92
Arterial HbO_2 (%)	94	92–95
Arterial carboxyhemoglobin (%)	4	4–5
Arterial pH	7.48	7.40–7.53
Arterial PCO_2 (mmHg)	40	35–44
Arterial plasma HCO_3 (mEq/L)	29	22–33
Arterial buffer base (mEq/L)	45	40–52

Modified from Swindle and Smith [15]

4 Notes

1. Other components and methods for sedation and anesthesia can be used. We use this mixture for its quick onset of anesthesia and the duration of sedation which is of high importance for a safe transportation to the operating facilities. A sufficient sedation is usually achieved after 10–15 min.
2. Weighting and clipping are performed in a separated room with the additional use of 2–3% of isoflurane in room air via facemask.
3. Successful intubation needs to be verified via auscultation of both lungs.
4. The catheter is used for repeated blood draws and infusion of fluids (Ringer's lactate solution, 5–10 mL/kg/h) during the entire experiment.
5. In our opinion, fentanyl patches are the most effective in both control and reduction of postoperative activity. This is in line

with previous literature and of high relevance in order to achieve successful long-term observations of wounds which is only possible with intact dressings [15].

6. Other methods to cause wounds in animals can be used. Another method was described by Middelkoop et al. They used an electrical dermatome with a depth of 2.7 mm [19].
7. The percentage of total surface body area (%TSBA) burned depends on the study protocol and the aims of the study. Most studies have not used a burn wound not greater than 40% TBSA [20]. TBSA for pigs is calculated using the validated Wachtel formula with known weight (TBSA ($m^2 = 0.121 \times$ wt $(kg)^{0.575}$) [21].
8. We stain the taken biopsies with hematoxylin and eosin and with Masson's trichrome to determine the extent of burn as determined by the depth of collagen degeneration. The described burn procedure leads to the complete coagulation of the superficial and most of the deep collagenous structures.
9. FCS, which contains protease inhibitors, can be added to better separate the two layers.
10. A cell count can be carried out with a hemocytometer. Cell viability may be assessed via trypan blue exclusion.
11. Instead of de-epidermized skin, transwell systems may be used for culturing the engineered skin construct [22].
12. Since pigs and other animals tend to repeatedly scratch their back, a multilayer dressing is necessary to protect the wounds especially in the first days after burn.
13. Since the use of second layer after day 5 frequently leads to hypergranulation and sporadic graft lost, we only use the second layer for the initial dressing application and the first dressing change.
14. Special swine clothing and goat tubes (intended to expand and provide a comfortable fit) may be positioned around the torso of the animal.
15. Each dressing change has to be made with full anesthesia and preferably with intubation. Mask anesthesia should only be used for quick dressing changes.
16. Note that donor sites are larger in area than individual wound sites; therefore, four punch biopsies may be obtained for each donor site, whereas three punch biopsies may be obtained for each burn wound site.
17. The angle to the wound should be as stable as possible using a 90° angle and the same distance at every time point.

References

1. Shevchenko RV, James SL, James SE (2010) A review of tissue-engineered skin bioconstructs available for skin reconstruction. J R Soc Interface 7:229–258
2. Huang S, Fu X (2011) Tissue-engineered skin: bottleneck or breakthrough. Int J Burn Trauma 1(1):1–10
3. Finnerty CC et al (2016) Hypertrophic scarring: the greatest unmet challenge following burn injury. Lancet 388(10052):1427–1436
4. Saaiq M, Zaib S, Ahmad S (2012) Early excision and grafting versus delayed excision and grating of deep thermal burns up to 40% total body surface area: a comparison of outcome. Ann Burns Fire Disasters 25(3):143–147
5. Ayaz M et al (2014) Early Excision and grafting versus delayed skin grafting in burns covering less than 15% of total body surface area; a non-randomized clinical trial. Bull Emerg Trauma 2 (4):141–145
6. Herndon DN et al (1989) A comparison of conservative versus early excision. Therapies in severely burned patients. Ann Surg 209 (5):547–553
7. Kapp MB (2006) Ethical and legal issues in research involving human subjects: do you want a piece of me? J Clin Pathol 59(4):335–339
8. Rheinwald JG, Green H (1977) Epidermal growth factor and the multiplication of cultured human epidermal keratinocytes. Nature 265(5593):421–424
9. Sullivan TP et al (2001) The pig as a model for human wound healing. Wound Repair Regen 9 (2):66–76
10. Ignacio G, El-Amin I, Mendenhall V (2016) Animal models for wound healing. in Skin tissue engineering and regenerative medicine eds. Albanna M, JH Holmes IV, New York: Academic Press/Elsevier 387–400
11. Meyer W, Schwarz R, Neurand K (1978) The skin of domestic mammals as a model for the human skin, with special reference to the domestic pig. Curr Probl Dermatol 7:39–52
12. Heinrich W et al (1971) Isolation and characterization of the large cyanogen bromide peptides from the alpha1- and alpha2-chains of pig skin collagen. FEBS Lett 16(1):63–67
13. Morris GM, Hopewell JW (1990) Epidermal cell kinetics of the pig: a review. Cell Tissue Kinet 23(4):271–282
14. National Research Council (1996) Guide for the care and use of laboratory animals. National Academy Press, Washington, DC. ISBN: 0309132622
15. Branski LK et al (2008) A porcine model of full-thickness burn, excision and skin autografting. Burns 34(8):1119–1127
16. Lammers G et al (2011) An overview of methods for the in vivo evaluation of tissue-engineered skin constructs. Tissue Eng 17 (1):33–55
17. Verhaegen PD et al (2009) Differences in collagen architecture between keloid, hypertrophic scar, normotrophic scar, and normal skin: an objective histopathological analysis. Wound Repair Regen 17(5):649–656
18. Swindle MM, Smith AC (2007) Swine in the laboratory: surgery, anesthesia, imaging, and experimental techniques, vol 2. CRC Press, Boca Raton, FL. ISBN: 9781466553477
19. Middelkoop E et al (2004) Porcine wound model for skin substitution and burn treatment. Biomaterials 25(9):1559–1567
20. Abdullahi A, Amini-Nik S, Jeschke MG (2014) Animal models in burn research. Cell Mol Life Sci 71(17):3241–3255
21. Wachtel TL, et al. (1972) Determining the surface areas of miniature swine and domestic swine by geometric design – a comparative study. USAARL Report, 73-3
22. Braziulis E et al (2011) Skingineering I: engineering porcine dermo-epidermal skin analogues for autologous transplantation in a large animal model. Pediatr Surg Int 27:241–247

Chapter 20

Transplantation of Autologous Dermo-Epidermal Skin Substitutes in a Pig Model

Thea Fleischmann, Flora Nicholls, Miriam Lipiski, Margarete Arras, and Nikola Cesarovic

Abstract

Due to its similarity of skin anatomy and physiology, the pig appears to be a well-suited animal model for preclinical studies of skin analog transplantations. The choice of the location of the skin defect and appropriate postoperative measures are essential for the protection of the transplanted graft. This protocol describes in detail a porcine skin transplantation model including peri- and postoperative measures taken to improve and refine the study outcome.

Key words Pig, Skin transplantation, Dermo-epidermal skin substitute, Perioperative refinement, Animal model

1 Introduction

Pigs appear to be the most concordant animal model for preclinical studies of skin analog transplantations for two main reasons [1]. First, pig skin is very similar to human skin in terms of its anatomical structure (e.g., similar dermal-epidermal thickness ratio, fixed subcutaneous layer, and hair follicles like humans) [2], its physiology, as well as its wound healing characteristics (e.g., in contrary to rodents, primarily by re-epithelialization rather than contraction) [3]. Secondly, the size of the pig allows transplantation of sizeable grafts equal to commercially available cultured epidermal autografts currently used in clinics [4].

The challenges of the model lay in the maximum protection of the skin graft during the first postoperative weeks. Choosing the optimal location for the skin defect is therefore crucial to prevent the animal from scratching the wound site with the mouth or the extremities. As pigs should be, whenever possible, housed in groups, protection of the skin graft from pen mates presents another challenge.

Sophie Böttcher-Haberzeth and Thomas Biedermann (eds.), *Skin Tissue Engineering: Methods and Protocols*, Methods in Molecular Biology, vol. 1993, https://doi.org/10.1007/978-1-4939-9473-1_20,

We here describe a porcine model, which permits the transplantation of skin analogous grafts measuring 7.5 × 7.5 cm [5]. An adequate pain management protocol, single housing of the animals, and a custom-made protective garment ensured the survival of the graft in the decisive initial phase. Additionally, to enhance the scientific readout, the use of a silicon chamber to compartmentalize the operation field from the surrounding skin proved to be of great value as it prevented healing from the wound edges themselves.

The protocol described here in detail should be of interest to veterinarians, veterinary technicians, animal caretakers, or other personnel responsible for the peri- and postoperative care of pigs involved in skin transplantation studies.

2 Materials

2.1 Animals

For the development of this protocol, domestic pigs, breed "Large White," aged between 2 and 3 months, weighing ca. 30–35 kg, were purchased from a local breeder. The animals were brought to the facility 2 weeks prior to the start of the experiment to allow appropriate acclimatization (*see* **Note 1**). Each pig was thoroughly examined within the acclimatization period to ensure that animals are healthy. Pigs were group-housed under standardized conditions: room temperature 19–22 °C, relative humidity 45–55%, and light cycle 12 h:12 h. The artificial light cycle was complemented by a natural light source from large skylights. Each animal received pelleted dry food twice daily and had ad libitum access to drinking water. Fruit yoghurt was given as a means of conditioning and reward (*see* **Note 2**). The general condition was checked twice daily throughout the study.

2.2 Veterinary Drugs

1. Anesthesia: Ketamine hydrochloride (Narketan®, 20 mg/kg BW), azaperone (Stresnil®, 1.25 mg/kg BW), atropine (0.03 mg/kg BW), isoflurane (Attane™, 1–2.5 Vol. %), propofol (Propofol-®Lipuro 1%, 1–2 mg/kg BW).
2. Sedation: Diazepam (Valium®, 0.2 mg/kg BW).
3. Analgesia: Buprenorphine (Temgesic®, 0.01 mg/kg BW), fentanyl patches (Durogesic® Matrix, 50 μg/h), carprofen (Rimadyl®, 3 mg/kg BW).
4. Antibiotics: Enrofloxacin (Baytril®, 2.5 mg/kg BW).
5. Antacid: Lansoprazole (Agopton® 30; 30 mg/day per os).
6. Wound treatment: Vita-Hexin®, Bepanthen® ointment.
7. Euthanasia: Pentobarbital (Eskonarkon®, 50 mg/kg BW).

2.3 Basic Surgical Materials

1. A sterile basic surgical kit including scissors, forceps, scalpel blades, scalpel holder, bowls, needle holder, swabs.
2. Fur clipper and disposable razors.
3. Sterile folio drapes (Foliodrape®).
4. Skin stapler, staple remover.
5. Suture material: 0-0 Ethilon, 1/0 Vicryl (Johnson & Johnson).
6. Skin disinfectant (Betadine®).
7. Sodium chloride solution: 0.9% NaCl.
8. Wound spray (Opsite®).

2.4 Materials for Skin Transplantation

1. Wound dressings: Silon-SES®, Acticoat (Smith & Nephew), Mepilex® Ag (Mölnlycke), Ligasano®, Moltex.
2. Transplantation chamber: Custom-made silicone consists of a chamber and a cover plate. The elastic chamber has a hardness (shore A) of 80 and the flexible cover has hardness (shore A) of 25. The molded silicone rubber parts are biocompatible, biostable, and heat resistant (200 °C) and thus amenable to standard sterilization processes, e.g., steam sterilization.
3. Protective suit: It is advisable to purchase a suit to protect the skin graft after transplantation. We obtained a made-to-measure suit with the following special features: Longitudinal and transversal stitching, Neoprene neck element, sufficient length on the ventral side, thick fleece padding around front and hind legs, and two lateral zippers divided into a cranial and a caudal part. At least two suits were manufactured for each animal in different sizes taking the growth of the animals into consideration (*see* Fig. 1).

2.5 Medium

1. Transport medium: 500 mL DMEM, 1 g/L D-glucose, 110 mg/L pyruvate, 15 mL 10,000 units/mL penicillin/10,000 μg/mL streptomycin, 0.75 mL gentamycin, 0.5 mL 250 μg/mL, amphotericin B.

3 Methods

3.1 Animal Preparation

1. Prior to the procedure, the pigs are fasted overnight to avoid vomiting and aspiration.
2. A deep sedation is performed in the stable with an intramuscular injection of ketamine hydrochloride (20 mg/kg BW), azaperone (1.25 mg/kg BW), and atropine (0.03 mg/kg BW). To ensure stress-free injection the pig is distracted during the injection with the above-mentioned rewarding food.

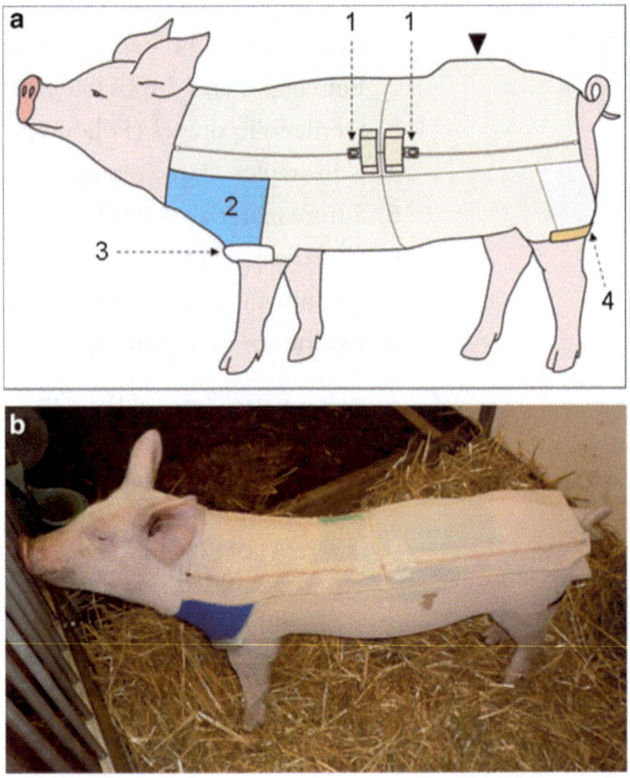

Fig. 1 Customized suit. (**a**) Schematic drawing of the customized garment for pigs, indicating the essential features: *1* two separate zippers, *2* neoprene insertion, *3* border of fleece upholstery, *4* elastic fabric strip. Arrowhead indicates location of shock-absorber device under the suit. (**b**) Picture showing the "custom-made" suit on a pig. Adapted from Ref. [5]

3. After loss of the righting reflex, a venous catheter is placed in the auricular vein and propofol is administered (bolus 1–2 mg/kg BW i.v.) to achieve relaxation and diminishment of the swallowing reflex for intubation.
4. An endotracheal tube (diameter 8 mm) is placed and anesthesia is maintained by inhalation anesthesia with isoflurane (1.0–2.5 Vol%) in 50:50 oxygen:air gas mixture (3–5 L/min) under positive-pressure ventilation.
5. Buprenorphine (0.01 mg/kg BW i.v.) is administered for perioperative pain management and enrofloxacin (2.5 mg/kg BW i.v.) as perioperative antibiosis.
6. Throughout the procedure, the animal should receive fluid supplementation with Ringer's lactate (10 mL/kg/h).
7. The pig is placed in dorsal recumbency on the operating table. Anesthetic monitoring is installed (pulse oximetry, temperature measurement, etCO_2, O_2, isoflurane, invasive blood pressure (A. coccygeal), ECG) and vital signs are monitored throughout the procedure.

3.2 Skin Biopsy Procedure

1. Shave and wash the surgical area with soap, and then perform three cycles of disinfection with Betadine®. Cover the surgical area with sterile drapes and neutralize the area with 0.9% NaCl.
2. The skin biopsy should be taken from the ventral abdomen, lateral and parallel to the mammary teats. Use a scalpel blade No. 10 and take a skin biopsy of about 15 cm length and 2 cm widths, total 15 cm^2 depth from epidermis to fascia.
3. Transfer the skin immediately into transport medium.
4. Close the wound in two layers, e.g., subcutaneous closure with resorbable Vicryl 1/0 and cutaneous wound closure with non-resorbable Ethilon 0/0 in single interrupted sutures.
5. Apply Opsite® wound spray and adhesive wound dressing.
6. After the intervention, anesthesia is discontinued and the pig brought back to the stable. For undisturbed recovery, the pig is kept individually in a single compartment on soft bedding under a warming lamp and is closely monitored until fully recovered.
7. The pigs receive 3 mg/kg BW carprofen i.v. postoperatively and a second dose of 0.01 mg/kg buprenorphine i.m. 6 h after the first application.
8. The animals are checked 2×/day for the following 3 weeks to ensure proper wound healing.

3.3 Skin Analog Transplantation

After the skin biopsy was taken, the pigs are granted a recovery time of 2–3 weeks. Ensure general well-being and appropriate wound healing by checking the animal regularly. During this period, each pig needs to be measured (body length, shoulder to elbow, hip to knee) so a made-to-measure protective suit can be manufactured.

3.4 Animal Preparation

1. Anesthesia and preparation of the animal are performed as described above for the skin biopsy.
2. The pig is then placed in sternal recumbency on the operating table. Anesthetic monitoring is installed (pulse oximetry, temperature measurement, $etCO_2$, O_2, isoflurane, invasive blood pressure (A. coccygeal), ECG) and vital signs are monitored throughout the procedure.

3.5 Skin Transplantation Procedure

1. A transplantation site was chosen medial to the vertebral column over the lumbar area (7.5 × 7.5 cm, square shaped) to minimize pain or discomfort during normal ambulation after the surgery.
2. Shave and wash the surgical area with soap and mark the implantation site with a waterproof pen. Perform three cycles of disinfection with Betadine® and cover the surgical area with sterile drapes.

3. Excise the skin in the marked area. The entire subcutaneous fat layer should be excised, but be careful not to injure the vascularized layer above the fascia.
4. Mobilize the skin edges to introduce the chamber. Cut the silicon edge of the chamber appropriately, until it fits smoothly into the skin.
5. Fix the chamber with 0-0 Ethilon on all sides.
6. Disinfect the wound inside the chamber with Betadine® and thoroughly neutralize with sterile 0.9% NaCl.
7. Apply the skin transplant inside the chamber and fix with skin clamps around the edges.
8. Apply silicon (Silon-SES®) directly on top of the transplant.
9. Apply Mepilex® Ag (Mölnlycke), 10 × 10 cm, and the silicon lid on top of the chamber.
10. Apply Mepilex® Ag (Mölnlycke), 20 × 20 cm, folio drapes, Ligasano®, 25 × 25 cm, folio drapes®, and Moltex (*see* **Note 3** and Fig. 2).

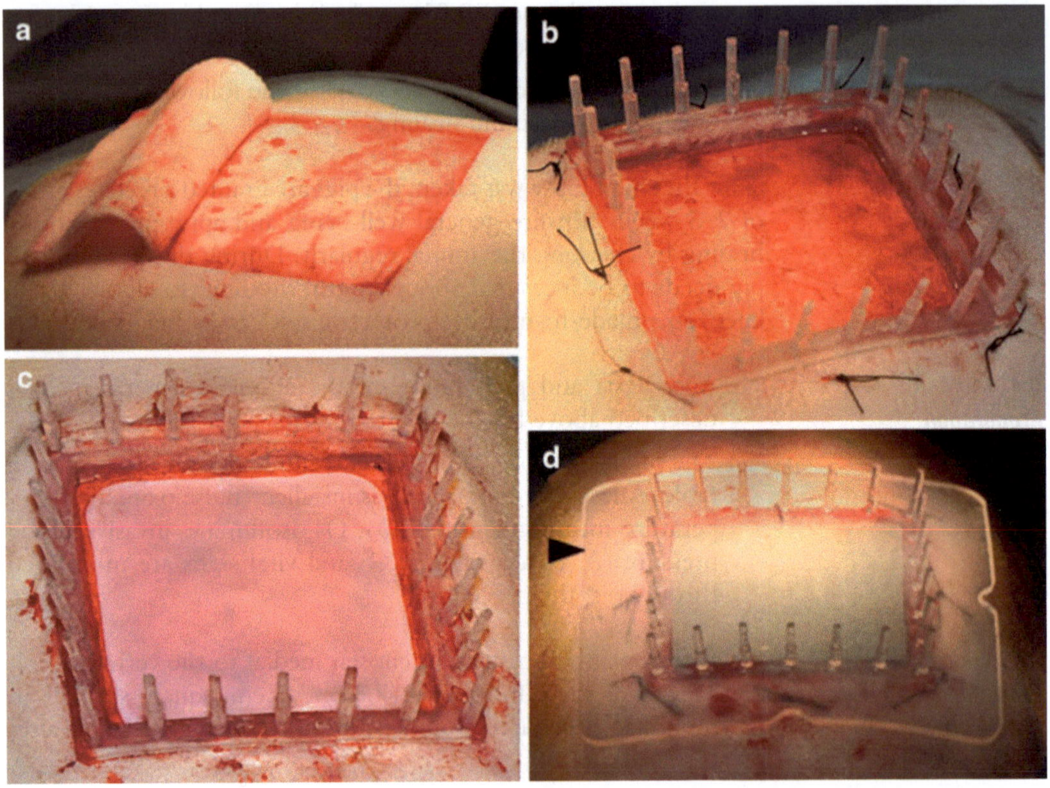

Fig. 2 Surgical procedure of skin substitute transplantation. (**a**) A full-thickness skin defect of 7.5 × 7.5 cm, located over the lumbar area, is created. (**b**) The silicone chamber is implanted and secured by fixation sutures. (**c**) The cultured skin substitute is transplanted onto the full-thickness wound. (**d**) The wound is covered with soft silver foam (gray) and a 2 mm thick silicone cover plate (arrowhead). Adapted from Ref. [5]

11. Finally, the made-to-measure suit should be put on while the animal is still anesthetized.
12. A fentanyl patch (Durogesic patch, 50 μg/h) is applied on the skin between the shoulder blades and fixed with adhesive tape. The patch is left in place for 3 days to provide continuous analgesia.
13. The pig is brought back to the stable to recover from anesthesia. Diazepam (0.2 mg/kg BW i.m.) is administered to ensure a smooth recovery and to reduce stress caused by the unfamiliarity of the protective suit. All animals receive diazepam (0.4 mg/kg BW; SID, p.o.) for 3 days to minimize agitation until the animals fully accustom to the suit.
14. Antibiosis is continued for 2 days after the intervention. To reduce the risk of stress-induced gastric ulcers, pigs receive lansoprazole (30 mg/day per os) for 5 days.

3.6 Post-operative Care

1. Pigs are checked twice daily for general health and well-being with emphasis on mobility, food and water intake, and signs of pain or wound infection and proper fitting of the suit. During the daily inspections, the suit is left in place to avoid disturbance to the healing of the transplanted skin graft.
2. A soothing cream (VitaHexin® or Bepanthen® ointment) should be applied under the armpits and groin on a regular basis to prevent irritation of the skin by the protective suit.
3. Pigs are now housed individually in neighboring sties to minimize the risk of peer-inflicted damage to the operation site (*see* **Notes 4** and **5**).

3.7 Inspection of Wound and Evaluation of Skin Analog

The transplant site is checked every 7 days following transplant surgery.

3.8 Animal Preparation

1. A deep sedation is performed in the stable with an intramuscular injection of ketamine hydrochloride (20 mg/kg BW), azaperone (1.25 mg/kg BW), and atropine (0.03 mg/kg BW).
2. After loss of righting reflex, a venous catheter is placed in the auricular vein and the animal is placed in sternal recumbency on the operating table.
3. Anesthesia is maintained by inhalation anesthesia with isoflurane (1.0–2.5 Vol%) in 50:50 oxygen:air gas mixture (3–5 L/min) under spontaneous breathing via a face-mask.
4. Anesthetic monitoring is installed (pulse oximetry, temperature measurement) and vital signs are monitored throughout the procedure.

3.9 Wound Inspection

1. Remove the suit and bandages and clean the wound site.
2. Open the silicon lid and disinfect the skin around the chamber with Betadine®.
3. Exchange Silon-SES® only when needed. Otherwise it should be left in place without further disinfection.
4. Apply chamber lid and bandage again.
5. Put suit back on.
6. Discontinue anesthesia and let the animal recover.
7. After each inspection of the transplant site, the made-to-measure suit is exchanged for a clean one while the dirty suit is being washed (*see* **Notes 6** and 7).

3.10 Euthanasia and Excision of Skin Analog

1. Anesthesia and preparation of the animal are performed as described above.
2. Remove the wound drapes and the silicon chamber.
3. Carefully excise the transplant and prepare it for histological examination.
4. Perform euthanasia with intravenous administration of pentobarbital (75 mg/kg BW).

4 Notes

1. It is important to handle the pigs daily during the acclimatization phase to familiarize the animals and to allow unstressed handling. It facilitates daily observations, clinical examination, or repositioning of the suit.
2. We offer our pigs some special treats; for example each pig daily receives some yogurt to make administration of anesthesia, clinical examinations (e.g., body temperature measurement), or readjustment of suits easier and stress free.
3. The several layers of dressing and the silicone chamber act as a "multilayer shock absorbent system," consisting of the silicon chamber and a 2 mm thick silicone cover plate to shield the area of interest from the wound edges; wound sites are covered with a soft silver foam dressing and fixation tapes encasing the foam, with additional foam and made-to-measure suit (*see* Fig. 3).
4. After putting on the suits, the pigs should be individually housed with visual and olfactory contact to fellow animals. Group housing was no longer possible, as the animals manipulated each other's suits until the suits were ripped apart or completely removed.
5. If possible, pigs should be acclimatized to the made-to-measure protective suits for a few days prior to the skin

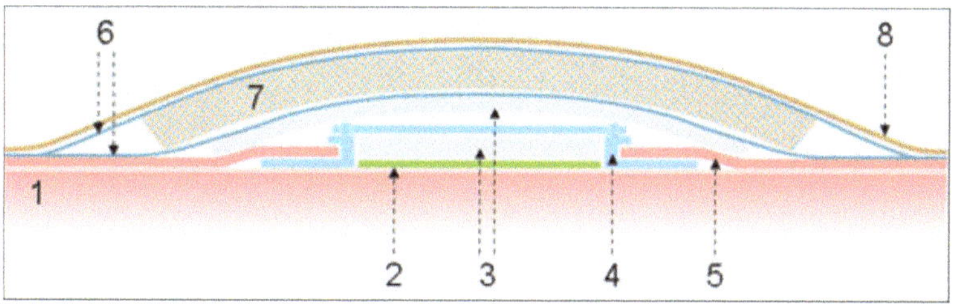

Fig. 3 Multilayered shock-absorber device. Schematic drawing showing the architecture of the device: *1* muscle including overlying fascia, *2* skin analog, *3* soft silver foam, *4* silicone chamber, *5* pig skin, *6* fixation tapes encasing the foam, *7* foam, *8* customized suit. Adapted from Ref. [5]

transplantation procedure. However, it might not be possible to put on the suit stress free in the conscious animal. In our study, we therefore put them on while the animal was still anesthetized, and carefully observed the animal during the recovery phase.

Diazepam (0.2 mg/kg BW i.m.) can be administered to support a smooth recovery and to reduce any stress due to the unfamiliarity of the suit. All animals further received diazepam (0.4 mg/kg BW; SID, p.o.) for 3 days to minimize agitation until the animals accustomed to the suit.

6. One should obtain at least two made-to-measure suits for each animal, taking the growth of the pig into consideration. We exchanged the suits during each inspection of the transplant site or in case of soiling and wetness.
7. Despite the fleece padding, we observed some redness under the armpits and groin. Therefore, a soothing cream (Vita-Hexin® or Bepanthen® ointment) should be applied on a regular basis.

References

1. Sullivan TP, Eaglstein WH, Davis SC, Mertz P (2001) The pig as a model for human wound healing. Wound Repair Regen 9(2):66–76
2. Lindblad WJ (2008) Considerations for selecting the correct animal model for dermal wound-healing studies. J Biomater Sci Polym Ed 19 (8):1087–1096
3. Swindle MM, Makin A, Herron AJ, Clubb FJ Jr, Frazier KS (2012) Swine as models in biomedical research and toxicology testing. Vet Pathol 49(2):344–356
4. Bottcher-Haberzeth S, Biedermann T, Reichmann E (2010) Tissue engineering of skin. Burns 36(4):450–460
5. Schiestl C, Biedermann T, Braziulis E, Hartmann-Fritsch F, Böttcher-Haberzeth S, Arras M, Cesarovic N, Nicolls F, Linti C, Reichmann E, Meuli M (2011) Skingineering II: transplantation of large-scale laboratory-grown skin analogues in a new pig model. Pediatr Surg Int 27:249–254

Index

Sophie Böttcher-Haberzeth and Thomas Biedermann (eds.), *Skin Tissue Engineering: Methods and Protocols*, Methods in Molecular Biology, vol. 1993, https://doi.org/10.1007/978-1-4939-9473-1,

MIX
Papier aus verantwortungsvollen Quellen
Paper from responsible sources
FSC® C105338

If you have any concerns about our products,
you can contact us on
ProductSafety@springernature.com

In case Publisher is established outside the EU,
the EU authorized representative is:
Springer Nature Customer Service Center GmbH
Europaplatz 3, 69115 Heidelberg, Germany

Printed by Libri Plureos GmbH
in Hamburg, Germany